Global Pesticide Resistance in Arthropods

GLOBAL PESTICIDE RESISTANCE IN ARTHROPODS

Edited by

M.E. Whalon

Professor, Department of Entomology, Michigan State University, East Lansing, Michigan, USA

D. Mota-Sanchez

Visiting Research Associate, Department of Entomology, Michigan State University, East Lansing, Michigan, USA

and

R.M. Hollingworth

Professor, Department of Entomology, Michigan State University, East Lansing, Michigan, USA

www.cabi.org

CABI is a trading name of CAB International

CABI Head Office
Nosworthy Way
Wallingford
Oxfordshire OX10 8DE
UK

Tel: +44 (0)1491 832111
Fax: +44 (0)1491 833508
E-mail: cabi@cabi.org
Website: www.cabi.org

CABI North American Office
875 Massachusetts Avenue
7th Floor
Cambridge, MA 02139
USA

Tel: +1 617 395 4056
Fax: +1 617 354 6875
E-mail: cabi-nao@cabi.org

A catalogue record for this book is available from the British Library, London, UK.

Library of Congress Cataloging-in-Publication Data
Global pestcide rersistance in arthropods / edited by M.E> Whalon, D.Mota-Sanchez, R.M. Hollingworth.
    p. cm.
  Includes bibliographical references and index.
  ISBN 978-1-84593-353-1 (alk. paper)
  1. Arthropod pests–Insecticide resistance. 2. Arthropod pests–Control. I. Whalon, Mark Edward, 1948- II. Mota-Sanchez, D. (David), 1960-III. Hollingworth, Robert M., 1939- IV. Title.

  SB951.5.G585 2008
  632'. 65–dc22

                                              2007040097

ISBN: 978 1 84593 353 1

Typeset by MRM Graphics Ltd, Winslow, UK.
Printed and bound in the UK by Cromwell Press, Trowbridge.

Contents

Contributing Authors

D.A. Andow, *Department of Entomology, University of Minnesota, 219 Hodson Hall, 1980 Folwell Ave., St Paul, MN 55108, USA.*

M.A. Caprio, *Department of Entomology and Plant Pathology, Mississippi State University, Mississippi State, MS 39762, USA.*

I. Denholm, *Rothamsted Research, Harpenden, Hertfordshire, AL5 2JG, UK.*

K. Dong, *Department of Entomology, Michigan State University, 106 Center for Integrated Plant Systems, East Lansing, MI 48824–1311, USA.*

G.P. Fitt, *CSIRO Long Pocket Laboratories, 120 Meiers Road, Indooroopilly, Queensland, 4068 Australia.*

E.J. Grafius, *Department of Entomology, Michigan State University, 442 Natural Science, East Lansing, MI 48824, USA.*

R.M. Hollingworth, *Department of Entomology, Michigan State University, 51 Natural Science, East Lansing, MI 48824, USA.*

R.E. Jackson, *USDA–ARS, 141 Experiment Station Road, Stoneville, MS 38776, USA.*

P. Leonard, *Regulatory and Government Affairs, BASF Belgium, Chaussée de la Hulpe 178, B – 1170 Brussels, Belgium.*

A.H.N. Maia, *Embrapa Meio Ambiente, PO Box 69, Jaguariuna, SP, Brazil.*

S. Matten, *USEPA Headquarters, Ariel Rios Building, 120 Pennsylvania Ave, NW, Mail Code: 7511P, Washington, DC 20460, USA.*

D. Mota-Sanchez, *Department of Entomology, Michigan State University, B17D Food Safety and Toxicology Center, East Lansing, MI 48824, USA.*

S. Peck, *Department of Biology, Brigham Young University, 155 WIDB Provo, UT 84602, USA.*

E.B. Radcliffe, *Department of Entomology, University of Minnesota, 219 Hudson Hall, 1980 Folwell Ave., St Paul, MN 55108, USA.*

D.W. Ragsdale, *Department of Entomology, University of Minnesota, 219 Hudson Hall, 1980 Folwell Ave., St Paul, MN 55108, USA.*

L. Rossiter, *NSW Department of Primary Industries, Australian Cotton Research Institute, Locked Bag 1000, Narrabi NSW 2390, Australia.*

M.S. Sisterson, *USDA–ARS, San Joaquin Valley Ag Sciences Center, 9611 S. Riverbend Ave., Parlier, CA 93648, USA.*

N. Storer, *Dow AgroSciences LLC, 9330 Zionsville Rd, Indianapolis, IN 46268–1054, USA.*

G.D. Thompson, *Dow AgroSciences LLC, 9330 Zionsville Rd, Indianapolis, IN 46268–1054, USA.*

M.E. Whalon, *Department of Entomology, Michigan State University, B11 Center for Integrated Plant Systems, East Lansing, MI 48824, USA.*

X. Qiang, *Office of Planning and Budgets, Michigan State University, 325 Hannah Administrative Building, East Lansing, MI 48824, USA.*

Introduction

MARK E. WHALON

Department of Entomology, Michigan State University, East Lansing, USA

Pesticide resistance within the order Arthropoda has been a subject of interest in the entomological sciences and among pest management professionals since A.L. Melander first asked the question, 'Can insects become resistant to sprays?' (Melander, 1914). He had observed high populations of the San Jose scale *Quadraspidiotus perniciosus* (Comstock), in lime sulphur-treated deciduous fruit. Following this report, published cases of resistance grew in frequency, probably reaching a high point in the late 1970s and early 1980s when many arthropod toxicologists, economic entomologists and evolutionary biologists busily reported cases, underlying mechanisms and the genetics of resistance. This trend has continued into the 21st century and has been reported in the online publication Arthropod Pesticide Resistance Database, the APRD (http://www.pesticideresistance.com), since the early 1990s.

Many pest management scientists believe that pesticide resistance is one of the principal problems facing crop production, human health, and animal protection. In our database developed at Michigan State University, we have more than 7747 cases of resistance with more than 331 insecticide compounds involved. From the estimated 10,000 arthropod pests, 553 species are reported with resistance to insecticides. Arthropod competition for food and niche space, and human and animal health concerns are the principal factors necessitating pest control. The intensity of pesticide use is linked to selection pressure, and this genetic selection pressure operating at the population level has repeatedly resulted in resistance and control failure across all of the common uses of insecticides and miticides globally.

Insecticide resistance history has demonstrated that control efforts increase with more stringent health and legal restrictions, as well as market-based quality or cosmetic standards. These ever-increasing standards result in lower pesticide treatment thresholds, which in turn lead to increased efforts to limit pest populations by inevitably increasing pesticide applications. Quite naturally, increased intensity of pest management leads to greater genetic selection pressure, which brings on more resistance issues. Some of these operational factors that contribute to increased selection pressure and resistance development in the global marketplace include export/import health and phytosanitary standards, and invasive species eradication projects, as well as global pandemics that threaten human and animal survival, like avian influenza.

Two major factors probably contribute more to the reduction of arthropod resistance than any others: (i) the development and marketing of new biological and chemical products with novel modes of action; and (ii) integrated pest management (IPM). The introduction of IPM in the 1970s probably slowed insecticide selection pressure and reduced the potential for resistance simply by diversifying the mortality mechanisms that targeted populations were exposed to. With the development of IPM, 'integrated resistance management' (IRM) or simply 'resistance management' became a significant amendment to IPM strategies that were implemented to avert, or at the least ameliorate, resistance development as a conscious goal of a comprehensive IPM programme.

In addition, we must remember the historical and geographical contexts in which individual resistance cases develop because, as a rule, broad management theories do not prevent resistance. Only tactics and tools employed in everyday pest management can effectively avoid resistance. Certainly in a major way, just the rotation of pesticide with different modes of action can slow resistance development, although there are always exceptions to every rule. Many of the chemistries providing resistance selection pressure in the past have been curtailed significantly in some countries or regions, but many, if not most, of these older chemistries are still being deployed somewhere on the globe. Therefore, one of the collateral effects of resistance development that we have begun to report in the APRD, and which can have a significant impact on both old and newer chemistries alike, is cross-resistance. As described in Chapter 3, cross-resistance is the phenomenon whereby an arthropod population develops resistance to a selecting compound or agent while simultaneously becoming resistant to a second compound or agent without having been exposed to this compound. Presumably, therefore, cross-resistance occurs primarily when a resistance mechanism imparts resistance to more than one compound. In some instances, newer chemistries may experience resistance problems from a previous history of selection from natural defensive compounds deployed by host plants or by man-made pesticides.

Therefore, it is useful to maintain resistance records of arthropod resistance case by case not only for evaluating IPM and IRM strategies, but also for inferring the likeliness of cross-resistance problems in the deployment of new pest management tools. In some of these instances, it is possible to deploy simultaneous discriminating dosages of insecticides known to be indicators of different modes of resistance to further characterize field populations. This emergent classification can help pest managers to develop more durable pest control programmes in the field.

In 1991, George Georghiou and Angel Lagunes-Tejada published a list of resistance cases from 1914 to 1989 (Georghiou and Lagunes-Tejada, 1991). The book won immediate respect and became a useful tool for pest managers, industry workers and even government workers interested in regulating insecticides. Following the publication of *The Occurrence of Resistance to Pesticides in Arthropods – an Index of Cases Reported Through 1989* and after the 1993 Entomology Society of America symposia addressing insecticide resistance, Dr Georghiou approached me with the proposal that Michigan State University should assume the maintenance and updating of his initial efforts. After due consideration, Robert Hollingworth, then Director of the Pesticide Research Center at Michigan State University, and I decided to embark on the process of documenting and commenting upon the evolution of resistance to pest management tools in agriculture, and human and animal health protection. David Mota-Sanchez joined these efforts in 1998, and today this team of scientists maintains and advances the collection, publication and dissemination of resistance cases for posterity.

The purpose of the present book is to update the global community on the status of arthropod resistance. Yet this book diverges from its predecessor's efforts (Georghiou and Lagunes-Tejada, 1991) in the following manner. First, the current effort updates and corrects the enumeration of resistance cases from 1914 to 2007. Secondly, the Georghiou and Lagunes-Tejada book covered field cases only, while our effort comprehensively updates refereed journal reports of field cases but also expands reporting to laboratory resistance cases as well. These latter cases may add utility in some instances where field resistance relates to reports from laboratory studies either prior to field or post-field resistance. The relatively recent development and deployment of transgenic crops with arthropod management measures may also require a similar effort in the future, but at this time is beyond the scope of our efforts reported herein.

Beyond just the enumeration of resistance cases and for the first time, we report the developments and data from our online case-reporting mechanism and linked database (http://www.pesticideresistance.com) to introduce a new mechanism of getting resistance case information to the relevant resistance managers the world over via the World Wide Web.

Chapter 1 reports the findings and analysis contained in this updated database. We also report resistance case analysis by chemical mode of action, top 20 most problematic arthropods, and decade analysis. We view this effort as the beginning of an automated era of pesticide resistance case reporting globally.

Chapter 2 provides a more detailed presentation of the 7747 cases of arthropod resistance. Cases are organized by species and include information on resistant compounds and geographic sites of resistance. Information from this chapter can be found online in the APRD (http://www.pesticideresistance.com).

Chapter 3 of this book describes the principal mechanisms of resistance that allow arthropods to escape control, as discovered through laboratory toxicological and molecular genetic studies. By reviewing the scientific literature of the last 10 years, the complexities of mechanisms of resistance are discussed, including acetylcholinesterase insensitivity, the kdr factor (knockdown resistance due to altered sodium channels), enhanced metabolism, reduced penetration, and target site insensitivity, among others. The molecular biology of the resistance is also briefly described.

In Chapter 4 we present a discussion of the principal features governing the evolution of resistance including mutation, gene flow, initial gene frequency, fitness, genetic drift in arthropod resistance and selection intensity while discussing spatially complex models and their role in pest management practices. Resistance management techniques are also reviewed in Chapter 5 where current examples of arthropod resistance management in the field are highlighted. Specifically, this chapter addresses any generalizations that can be made about IPM and, more importantly, considers the barriers to field implementation of resistance management techniques and suggests further research to improve IPM strategies, tactics, and tools. Moreover, a portion of the chapter summarizes the use of transgenic *Bt* crops and their role in IPM.

Finally, Chapter 6 discusses historic attempts at resistance management policy and their modern counterparts in order to identify key strengths and weaknesses that must be addressed in future policy initiatives. Recent worldwide developments in pesticide policy, like the Food Quality Protection Act (FQPA) in the USA, and the manner in which these regulations affect the use of pesticides, IPM, and trade are evaluated comparatively. In addition, the chapter also addresses the role of institutions around the world responsible for insecticide resistance management including the international Insecticide Resistance Action Committee (IRAC), the United States Department of Agriculture Science and Management of Pesticide Resistance (WERA-60), and other efforts from across the globe.

References

Georghiou, G.P. and Lagunes-Tejada, A. (1991) *The Occurrence of Resistance to Pesticides in Arthropods. An Index of Cases Reported Through 1989.* Food and Agriculture Organization of the United Nations, Rome.

Melander, A.L. (1914) Can insects become resistant to sprays? *Journal of Economic Entomology* 7, 167–173.

1 Analysis of Global Pesticide Resistance in Arthropods

M.E. WHALON, D. MOTA-SANCHEZ AND R.M. HOLLINGWORTH

Department of Entomology, Michigan State University, East Lansing, Michigan, USA

1.1. Introduction

Most pest management scientists, as well as pesticide industry and government regulatory workers, agree that arthropod resistance is a very important driver of change in modern human health, agriculture, animal production, right of way management, and structural and urban pest management. There are many examples of health protection and agricultural production systems that have been vulnerable to the development of, and sometimes devastated by, the effects of arthropod resistance. For instance, early estimates of pesticide resistance impacts on crop protection in the USA exceeded $4 billion annually (Pimentel *et al.*, 1991). Later impact estimates on crop protection suggested a more modest, but nevertheless substantial impact of $1.4 billion annually in the USA (Pimentel *et al.*, 1993). If the impact of resistance is in the range of $1–4 billion in the USA, what must it be in the rest of the developed and developing world? Without better record keeping and resistance severity assessments, we may never know the answer to this question. Therefore, an overall objective of this chapter is to begin to address the question of the importance of arthropod resistance by examining the scientific literature's published cases of arthropod resistance globally.

While reporting arthropod resistance globally is an ambitious and incessant undertaking, it is vital none the less to those responsible for the amelioration or management of arthropod resistance development, thereby reducing the potentially devastating impacts of resistance on societies around the world. After all, the old saying 'what gets measured gets managed' has great relevance in policy development, resource allocation, and society's awareness of any significant development. Therefore, the building of a database that can receive resistance case reports over the World Wide Web while nearly instantaneously updating and reporting the world's arthropod resistance status is a necessary and significant objective with potential impacts on societies around the globe. This chapter reports the current status of arthropod resistance worldwide as contained and reported in the Arthropod Pesticide Resistance Database (APRD) housed at Michigan State University, East Lansing, Michigan, USA.

Various factors contribute to the development of arthropod resistance to insecticides and miticides including arthropod host and reproduction ecology, pesticide use intensity, frequency of exposure, pre-existing resistance alleles in treated populations, dose–toxicity relationships, and cross- and multiple-resistance phenomena, as well as societal factors including cosmetic appearance standards, entomophobia, phytosanitary trade laws, and a range of Integrated Pest Management (IPM) strategies, tactics, and tools. Properly understood and positioned in the lexicon of pest management, arthropod resistance management is an IPM strategy with an array of tactics and tools, including the current status of arthropod resistance reporting from the APRD (http://www.pesticideresistance.com).

The impact of pesticide resistance on human health is likely to be even more important globally than its agriculturally related effects. However, like agricultural resistance, many cases of arthropod resistance affecting human health are unreported due to the lack of infrastructure, resources, and technical personnel. Where case data are available, they reside most significantly in both national and international health organizations and, therefore, are not readily available. For example, mosquito resistance to insecticides is a principal fact of daily life, whereby the incidence of malaria has increased in many developing countries. Annually, malaria affects over 500 million people and kills 3 million. This can be attributed to the growing prevalence of insecticide-resistant mosquitoes (Roll Back Malaria Partnership, 2005; African Insect Science for Food and Health, 2007). Considering the number of arthropod-borne diseases, the number of arthropod vectors, and the frequency of arthropod resistance, controlling resistance for the sake of human and animal health is a daunting, but vital, challenge.

This chapter reports the next significant step in what is an evolving history of the development of systems designed to support arthropod resistance case reporting. These processes began in the USA and EU and have spread through international trade relations and health organizations throughout the world. Certainly, the authors are thankful for the contributions of organizations such as the International Service for National Agricultural Research (ISNAR) (www.isnar.cgiar.org) for their policy efforts with genetically engineered plant policy, which includes a Genetically Modified Resistance Management Plan and Process for any government interested in this information. We also acknowledge the leadership and commitment of the international Insecticide Resistance Action Committee (IRAC) (www.irac-online.org). These action committees have worked on various aspects of resistance management, including detection and monitoring programmes, yet IRAC's greatest development to date is probably the effort to develop resistance reporting by mode of action (MOA) classification of insecticides and miticides (Table 1.1). In this chapter, we use the IRAC MOA classes exclusively to describe insecticide resistance data. Our aim is to familiarize readers with the MOA groupings, as well as to encourage homogeneous reporting of future resistance cases.

Table 1.1. IRAC mode of action groups.

Group no.		Group name	Included chemistries
1		**Acetylcholinesterase inhibitors**	
	A	Carbamates	Aldicarb, benfuracarb, carbaryl, carbofuran, carbosulfan, fenobucarb, methiocarb, oxamyl, thiodicarb, triazamate
	B	Organophosphates	Acephate, chlorpyrifos, diazinon, dimethoate, malathion, methamidophos, monocrotophos, parathion-methyl, profenofos, terbufos
2		**GABA-gated chloride-channel antagonists**	
	A	Cyclodiene organochlorines	
	B	Phenylpyrazoles (fiproles)	
3		**Sodium-channel modulators**	
		Pyrethroids	Bifenthrin, cyfluthrin, cypermethrin, alpha-cypermethrin, zeta-cypermethrin, deltamethrin, DDT, esfenvalerate, etofenprox, lambda-cyhalothrin, methoxychlor, pyrethrins (pyrethrum), tefluthrin
4		**Nicotinic acetylcholine receptor agonists/antagonists**	
	A	Neonicotinoids	Acetamiprid, clothianidin, dinotefuran, imidacloprid, nitenpyram, thiacloprid, thiamethoxam
	B	Nicotine	Nicotine
	C	Bensultap	Bensultap
		Cartap	Cartap hydrochloride
		Nereistoxin analogues	Thiocyclam, thiosultap-sodium
5		**Nicotinic acetylcholine receptor agonists (not 4)**	
		Spinosyns	Spinosad
6		**Chloride-channel activators**	
		Avermectins and Milbemycins	Abamectin, emamectin benzoate, milbemectin
7		**Juvenile hormone mimics**	
	A	Juvenile hormone analogues	Hydroprene, kinoprene, methoprene
	B	Fenoxycarb	Fenoxycarb
	C	Pyriproxyfen	Pyriproxyfen
8		**Unknown or non-specific – fumigants**	
	A	Alkyl halides	Methylbromide
	B	Chloropicrin	Chloropicrin
	C	Sulphuryl fluoride	Sulphuryl fluoride
9		**Unknown or non-specific – selective feeding blockers**	
	A	Cryolite	Cryolite
	B	Pymetrozine	Pymetrozine
	C	Flonicamid	Flonicamid
10		**Unknown or non-specific – mite growth inhibitors**	
	A	Clofentezine	Clofentezine
		Hexythiazox	Hexythiazox
	B	Etoxazole	Etoxazole

Table 1.1. IRAC mode of action groups (*continued*).

Group no.	Group name	Included chemistries
11	**Microbial disruptors of insect midgut membranes – includes transgenic crops expressing** *Bacillus thuringiensis* **toxins,** *Bt* **subspecies shown**	
A1		*Bt israelensis*
A2		*Bacillus sphaericus*
B1		*Bt azawai*
B2		*Bt kurstaki*
C		*Bt tenebrionensis*
12	**Inhibitors of oxidative phosphorylation, disruptors of ATP formation (inhibitors of ATP synthase)**	
A	Diafenthiurion	Diafenthiurion
B	Organotin miticides	Azocyclotin, cyhexatin, fenbutatin oxide
C	Propargite	Propargite
	Tetradifon	Tetradifon
13	**Uncouplers of oxidative phosphorylation via disruption of H proton gradient**	
		Chlorfenapyr, DNOC
14	**Unallocated**	
15	**Inhibitors of chitin biosynthesis, type 0, Lepidopteran**	
	Benzoylureas	Chlorfluazuron, diflubenzuron, fluazuron, flucycloxuron, flufenoxuron, hexaflumuron, lufenuron, novaluron, noviflumuron, teflubenzuron, triflumuron
16	**Inhibitors of chitin biosynthesis, type 1, Homopteran**	
		Buprofezin
17	**Moulting disruptor, Dipteran**	
		Cyromazine
18	**Ecdysone agonists moulting disruptors**	
A	Diacylhydrazines	Chromafenozide, halofenozide, methoxyfenozide, tebufenozide
B		Azadirachtin
19	**Octopaminergic agonists**	
		Amitraz
20	**Coupling site II electron transport inhibitors (complex III)**	
A		Hydramethylnon
B		Acequinocyl
C		Fluacrypyrim
21	**Coupling site I electron transport inhibitors**	
	METI acaricides and Rotenone	Fenazaquin, fenpyroximate, pyrimidifen, tebufenpyrad, tolfenpyrad, rotenone
22	**Voltage-dependent sodium-channel blockers**	
		Indoxacarb
23	**Inhibitors of lipid synthesis**	
	Tetronic acid derivatives	Spirodiclofen, spiromesifen

Table 1.1. IRAC mode of action groups (*continued*).

Group no.	Group name	Included chemistries
24	**Mitochondrial complex IV electron transport inhibitors**	
A		Phosphides
B		Cyanide
C		Phosphine
25	**Neuronal inhibitors (unknown mode of action)**	
		Bifenazate
26	**Aconitase inhibitors**	
		Fluoroacetate
27	**Synergists**	
A	P450-dependent monooxygenase inhibitors	Piperonyl butoxide
B	Esterase inhibitors	DEF
28	**Ryanodine receptor modulator**	
		Flubendiamide
UN	**Compounds with unknown mode of action**	
A		Benzoximate
B		Chinomethionat
C		Dicofol
D		Pyridalyl
NS	**Miscellaneous non-specific (multi-site) inhibitors**	
A		Borax
B		Tartar emetic

IRAC, Insecticide Resistance Action Committee.

International health, food, and agriculture organizations, like the Centers for Disease Control and Prevention (CDC) (http://www.cdc.org), the World Health Organization (WHO) (www.who.int/en), and the Food and Agriculture Organization of the United Nations (FAO) (www.fao.org), have also contributed to the developments in arthropod resistance case reporting mentioned herein. With the development of an internationally accessible case-reporting mechanism as outlined below, we hope that arthropod resistance reporting will become even more accessible across these organizations and societies, becoming more reliable and useful for a multitude of purposes and users. By creating an online mechanism to report resistance incidence and management, perhaps societies the world over can share resistance management strategies, tactics, and tools to promote better agricultural production, human health, international cooperation, and scientific progress.

In this chapter, our intent is to report an analysis of the development and occurrence of arthropod resistance to insecticides and miticides by: (i) taxonomic classification (family, order, class); (ii) IRAC MOA; and (iii) date reported (since 1914 to present). The cases reported herein are all derived from APRD (http://www.pesticideresistance.com) and primarily reflect refereed journal reports of both field and laboratory resistance. Readers should understand that many factors contribute to the publication and enumeration of resistance events including academic and field worker interests, resources for monitoring resistance, pest identification resources, experience and capability of samplers, institutional structures that support formal pest management organizations with resistance expertise, and local and national laws stipulating resistance monitoring and reporting (see Mota-Sanchez *et al.*, Chapter 2, this volume). Thus the APRD contains resistance case reports over a range of importance classifications or severity levels, from none or low importance to high or extreme human suffering, loss of life, or grave economic loss. Although we are not reporting the resistance case severity in this chapter, the APRD is now archiving resistance severity level as a new category in the database. We anticipate that this severity rating will be increasingly important to pesticide producers, pest managers, regulators, academics, and the scientific press in the future. Another feature of the APRD is that resistance case reports from this site will be searchable by geopolitical region. More recently, with the development of Internet-based resistance case submissions, it is hoped that online arthropod resistance case submissions may enable and accelerate the advent of near real-time reporting of resistance for management locally and regionally. Submitted resistance cases are reviewed by experts from industry, academia, and applied IPM. What is available now is the broad dissemination of resistance case information, which can inform others in different settings throughout the global community, heightening the likelihood that resistance monitoring and reporting may lead to better resistance management. Our long-term hope is to add near real-time spatial reporting so that resistance management can become much more strategic and timely from a local, regional, and even global perspective. If you are interested in submitting a resistance case report to the APRD, we encourage you to go online to http://www.pesticideresistance.com and follow the instructions.

1.2. Definitions of Resistance

Resistance is the microevolutionary process whereby genetic adaptation through pesticide selection results in populations of arthropods which present unique and often more difficult management challenges (Whalon and McGaughey, 1998). One consequence of resistance is the failure of plant protection tools, tactics, or strategies to limit pest populations below economic injury levels where such failure is due to a genetic adaptation in the pest. This definition has traditionally been applied to insect populations that escape the effects of a chemical insecticide. Yet nearly all classes of organisms, but especially arthropods, have been submitted to extensive human-induced selection pressure, thus providing an array of examples of resistance microevolution to pest management measures, chemical or otherwise.

In the same way that resistance develops over time, the definition of resistance has also been adapted and refined over the years. A panel of WHO experts defined resistance as 'the development of an ability in a strain of insects to tolerate doses of toxicants which would prove lethal to the majority of individuals in a normal population of the same species' (WHO, 1957). This definition was the established operational definition for years, and it has been cited repeatedly. However, after more than 70 or 80 years of synthetic insecticide application in the world, arthropod populations have been exposed to and selected by one or more pesticides, significantly intensifying the difficulty in locating an unselected, or susceptible, population. The WHO definition reflects a population view rather than a focus on individual arthropods, which today must be expanded to individual testing and monitoring since biochemical and molecular genetic resistance detection techniques can be applied to single individuals. Screening for low-frequency resistance alleles, which is especially important for plant biopesticides and genetically modified organisms involving *Bacillus thuringiensis* (*Bt*) toxin-producing crops, would not fit the WHO definition cited above.

A more flexible definition was proposed by J.F. Crow that expanded consideration to the survival of single individuals within populations, as well as metapopulations. He proposed that 'resistance marks a genetic change in response to selection' (Crow, 1960). Because his definition was not limited to relatively high resistance expression levels, it is more useful in that it did not rest upon field failure of an insecticide, encouraging a management perspective by allowing for early detection. Thus, incipient resistance was included in his definition, paving the way for early detection and resistance management before the product in question failed. In essence, Crow introduced the 'modern era' of resistance management by foreseeing the need to manage resistance before it developed. That is, most tactics and strategies in resistance management can, and often are, introduced a priori to prevent resistance development rather than relying on detection or field failure to initiate countermeasures. Essentially, this has been the approach taken by the US Environmental Protection Agency (US EPA) in registering genetically modified field crops and other plant species which produce *Bt* toxins (see Thompson *et al.*, Chapter 6, this volume).

In 1987, R.M. Sawicki modified Crow's work by adding his own understanding of the genetics of selection. 'Resistance marks a genetic change in response to selection by toxicants that may impair control in the field' (Sawicki, 1987). In this view, Sawicki was careful to consider the possibility that resistance development may not economically impair control of the organism in real-world applications. In this approach, strains of organisms that are selected for pesticide resistance in the laboratory are therefore considered resistant. Today, a rich array of laboratory-selected strains adorn the scientific literature, and their evolution has led to unending discussion, and even heated arguments, about the value of such work when laboratory-selected strains may not reflect even the likely resistance development to a given selecting agent when field-scale populations are considered. However, we have also considered in the database cases of resistance developed in the laboratory, as they are important demonstrations of the potential development of resistance in the field and also important models to study the genetics and mechanisms of resistance.

For its part, the agrochemical industry has often led resistance science in the effort to understand, define, monitor, and manage pesticide resistance (www.irac-online.org). IRAC and other resistance action committees were formed by the pesticide industry after scientific, public, and new regulatory pressures arose following the exponential increase in the worldwide cases of resist-ance during the last half of the 20th century (and also the recognition by industry that new chemistries with novel modes of action are a precious resource that should be conserved). The criteria developed by IRAC for defining resistance are much more restricted than those described above and include the following cir-cumstances (Tomlin, 1997).

An insect should be viewed as resistant only when:

- The product for which resistance is being claimed carries a use recommen-dation against the particular pest mentioned, and has a history of successful performance.
- Product failure is not a consequence of incorrect storage, dilution, or appli-cation and is not due to unusual climatic or environmental conditions.
- The recommended dosages fail to suppress the pest populations below an economic threshold.
- Failure to control is due to a heritable change in the susceptibility of the pest populations to the product.

Based on these criteria, IRAC has repeatedly pointed out that the term 'resis-tance' should only be used in application to field failure, and only when field failure is confirmed scientifically. Although the IRAC criteria were sufficient to ensure that a pest population had truly developed resistance, this approach is still problematic since the genetic basis of the term 'resistance' is most effectively managed either by the assumption that resistance will develop or through early detection and subsequent deployment of resistance management tactics and tools. For the most part, early detection of low frequencies of resistant alleles in a population may not warrant a cry of resistance, yet it is a harbinger that re-sistance-mitigating tactics may be worthwhile where the longevity and utility of a company's product is at risk. The 'Achilles heel' of this approach to resistance management is that economic and proprietary interests of narrowly defining and delaying the acceptance of resistance may outweigh maintaining the long-term utility of the product to a sector of agriculture or human health protection. Thus a purely economic argument that determines that the establishment of re-sistance could preclude future use of a pest management tool, a class of tools, or worse, promote cross-resistant genes to currently undiscovered or unutilized tools. Thus the indiscriminate use of proprietary chemistries and genes has broader potential consequences and could lead to outcomes that are suboptimal for society as a whole.

1.3. Counting Resistant Arthropods

The remainder of this chapter updates and corrects the enumeration of resistance cases from 1914 to 2007 (Table 1.2). The basis of this enumeration is over 1694 refereed journal articles contained in the APRD (http://www. pes-ticideresistance.com). These extracted data reveal both the status of laboratory- and field-selected arthropod resistance in the published literature

Table 1.2. Case and species numbers of resistant arthropods by importance.

	Cases						Species					
	Total	Agricultural	Medical	Parasitoid	Other	Pollinator	Total	Agricultural	Medical	Parasitoid	Other	Pollinator
Acari	1025	708	248	69			76	41	23	12		
Araneae	1		1				1		1			
Coleoptera	884	860	13	8	3		74	68	1	3	2	
Copepoda	1		1				1		1			
Dermaptera	222	220			2		4		3		1	
Diptera	2265	291	1937	4	33		187	26	149	2	10	
Ephemeroptera	2				2		2				2	
Hemiptera	162	85	77				22	17	5			
Homoptera	992	989	3				58	57	1			
Hymenoptera	37	6	2	26		3	16	3	1	11		1
Lepidoptera	1799	1799					85	85				
Neuroptera	21			21			1			1		
Phthiraptera	114	4	110				9	1	8			
Siphonaptera	89		89				9		9			
Thysanoptera	133	133					8		8			
Total	7747	4875	2701	128	40	3	553	306	202	29	15	1
%	100	62.9	34.9	1.7	<1	<1	100	55.3	36.5	5.2	2.7	<1

globally as well as those data submitted through the online framework and review process, which has been available since February 2007 at the same web site (see Mota-Sanchez *et al.*, Chapter 2, this volume).

R. Busvine, an early pioneer in resistance case reporting, published a number of times in the *Bulletin of the WHO* (FAO, 1980). Following Busvine's early work, A.W.A. Brown published a series of resistance case tables for the WHO and other agencies beginning in the 1950s until the early 1970s. These early resistance reporters focused on human and animal disease vectors, which were some of the first targets of worldwide pesticide suppression (Brown, 1958). In the 1980s, Karen Theiling, a Master of Science student under the direction of Brian Croft at Oregon State University, began to collect and report cases of arthropod natural enemy resistance (Theiling and Croft, 1988). The thrust of this work was the culmination of Croft's development of the concepts surrounding 'pesticide selectivity,' which in some instances was based upon insecticide or acaricide resistance as an advantage in IPM programmes (Croft, 1990). The SELECTV database (http://ipmnet.org/phosure/database/selctv/selctv.htm) persists today at Oregon State University under the direction of Dr Paul Jepson.

1.4. Database Data, Information Extraction, and Summarization

The resistance incident report of Georghiou and Lagunes-Tejeda (1991) was the world's first effort to archive resistance incidents worldwide. Their book's data were a 'soft copy' enumeration of resistance cases filed in office cabinets, and an early database was later attempted. These resistance reports and articles were not readily available to the public even after the publication of the book. The library was located at the University of California Riverside campus and was inaccessible and very difficult and time-consuming to query. Despite all its shortcomings, the Georghiou and Lagunes-Tejeda text became a classic in the entomological and pesticide chemistry disciplines. Following the book's success and an increasing demand for updated resistance information, an interdisciplinary pesticide industry group, which evolved into IRAC, and an academic group (WRCC-60) under the aegis of the US Department of Agriculture (USDA) convened working groups with the intent of producing information that purported better stewardship of valuable insecticide tools. As a member of the latter and in communication with IRAC, Dr Mark Whalon at Michigan State University proposed a US interregional collaboration in the maintenance, expansion, and development of the resistance case literature. This concept and process continue today in both industry via IRAC and academia through WERA-1222. These two groups have collaborated extensively over the years and both share in the responsibility for the formation of the current APRD (http://www.pesticide resistance.com) and its accompanying *Resistant Pest Management Newsletter* (http://whalonlab.msu.edu/rpm/index.html). The development of information science software with powerful new database abilities and computer hardware with large information storage has allowed the current APRD to have many features that make access, query, and manipulation of this information possible. With the proliferation in accessibility and usage of the World Wide Web

(Townsend, 2001), these tools are now available to anyone on the planet with a computer and Internet access (see Mota-Sanchez *et al.*, Chapter 2, this volume). The tables in the present chapter have been extracted from the latest information derived from the APRD as of August 2007 (http://www.pesticideresistance.com).

However, there may be as many ways as there are resistance authors to observe and document a pesticide-resistant population. Despite this, standardized methods for resistance detection do exist. In fact, the FAO has been publishing standardized testing protocols for species affecting human health since 1969 (FAO, 1980). While continually improving laboratory techniques help to document resistance, authors still may interpret and report results of standardized tests differently. While there remains no established standard for reporting resistance, the authors would argue that the APRD may be approaching the establishment of such a standard (http://www.pesticideresistance.com). Certainly there are many factors in using the term 'resistant' pest, strain, or population that might have caused misunderstanding in many instances. Therefore, our strategy has been to rely upon the expertise of qualified scientific reviewers, leaving the ultimate decision for inclusion of a resistance case within the database to rest with the editorial board of the APRD researchers Drs Whalon, Hollingworth, and Mota-Sanchez. How 'resistance case' is defined is a critical point in documenting and recording resistance (see Chapter 2 to review the definition of a case of resistance and combinations of points and events that can complicate this definition). The editors believe that our approach exhibits agreement with other working definitions of resistance from the authors reviewed briefly above. Factors involved in deciphering a resistance report include the Whalon and McGaughey definition of resistance (Whalon and McGaughey, 1998), several intrinsic and extrinsic factors of the test itself (Busvine, 1968), and the type of statistical analyses used to report the population's resistance level (Ffrench-Constant and Roush, 1990).

These efforts were aimed to implement high quality and reliability standards in order to circumvent, or at least minimize, criticism of the data contained in the APRD. As a consequence of these methodologies, we believe the APRD probably under-reports legitimate resistance cases (see Mota-Sanchez *et al.*, Chapter 2, this volume). We would appreciate notification if readers find instances where the APRD has under-reported or missed a viable resistance case. We view the enumeration of resistant arthropods as a dynamic process; not only as new resistant populations develop, but also as past reports from around the world are uncovered, analysed and published in the APRD.

1.5. Resistance by Taxonomic Classification

Nearly a century ago, resistance in the San Jose scale *Quadrispidiotus perniciosus* (Comstock) to lime sulphur in deciduous fruits in the state of Washington, USA was the first documented case of a pesticide-resistant arthropod (Melander, 1914). As of August 2007, the current number of cases and species in the world is 7747 and 553, respectively (Fig. 1.1). The data from the APRD show that while there has been a dramatic 70% increase in the number of cases since Georghiou and Lagunes-Tejeda published their case list in 1991 (4553 cases and 503 species), the increase in the number of species showing resistance for the first time is only 9.9%.

The order with the highest number of resistant species is Diptera with 33.8%. Most of the resistant species in this order are vectors of human and animal diseases. The intense use of insecticide to protect against these insects resulted in this high selection of resistant species. Other major orders are Lepidoptera (15.4%), Acari (13.7%), Coleoptera (13.4%), Homoptera (10.5%), and Hemiptera (4.0%). Most of the species in these orders have been selected in an agricultural arena. However, there is an important number of species (76) in the order Acari that also have medical and veterinarian importance. Despite the high percentages of resistant species that have developed from the orders mentioned above, we cannot underestimate the importance of a few species from other orders. For instance, in Dermaptera, the German cockroach, *Blatella germanica*, has developed resistance to 42 different compounds (Table 1.3), while the head louse, *Pediculus humanus capitis*, presents many difficult resistance-related problems to pediculocide control in a relatively small insect order (Downs *et al.*, 2002).

1.6. Top Twenty Resistance Arthropods

One of the most frequently cited tabular outputs of the database is the top 20 ranked resistant arthropod pests in the world. Of course, some of these top 20 arthropods may not be a major resistance threat at specific locations, as this list reflects an accumulation of reports based on the number of unique cases

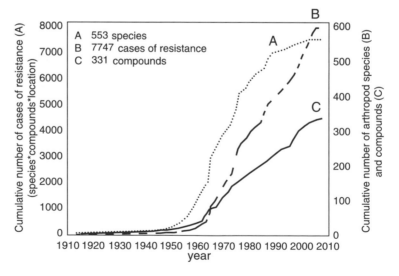

Fig. 1.1. A century of evolution of arthropod insecticide resistance.

involving a compound to which a documented resistance population was selected by its use at a specific location. Table 1.3 reports the current top 20 resistant arthropods based on this ranking. The list proffers arguably the 20 most difficult to manage and economically important arthropod pests anywhere. With new resistance cases reported steadily from 1943 to the present, most of these species would actually be included in this list were it compiled in any decade from the 1940s until now.

This last statement is a remarkable biological observation in the face of the very significant economic and health challenges that other pests mount across the globe. It is even more remarkable when one views the range of new and old resistance avoidance strategies, tactics and tools available today for pest managers. In many of these instances, pest managers or those who have made treatment decisions are probably choosing a single tactic, usually insecticide suppression, to limit the economic damage of these pests. The recent development of novel chemical modes of action, the application of molecular biology in the understanding of resistance evolution, and the development and expansion of the use of biological control agents and pheromones, together with the availability of botanical insecticides and repellents, gives ample strategies, tactics and tools to avoid creating new resistance cases in these 20 champion species. Yet the predominance of a single tactic, the strategy of population suppression, continues to promote resistance despite available alternatives for practising sound IPM.

What can we learn from this list? Probably one of the principal messages is that many of the other 533 pesticide-resistant species share a great deal of the genetic, biological and operational factors that together lead to resistance in species after species. Certainly as a science, pest management should be employing a range of resistance management strategies, tactics and tools. Diversifying the mortality mechanism by rotating modes of action, introducing cultural and biological controls, using discriminating dosage field monitoring tools, etc. could combine to give producers and pest management personnel critical tools to avoid resistance. Those who work with producers and their advisors have an obligation to educate and report the need for resistance management principles, as well as the consequences of retaining the single tactic of insecticidal population suppression as the primary method of controlling pests. The 'spray and spray' strategy utilizing a single active ingredient approach to pest control will no doubt continue to result in accelerated resistance development.

We stress, however, that the failure to achieve membership in the elite 20 most resistant insects and mites does not indicate that the status of a particular arthropod's resistance development is not important in an environmental, ecological, economic or social sense, especially in a local situation. Indeed, we believe that every case of resistance is important and should be observed, understood in context, reported, and used to teach better management, thereby improving human existence and the environment. Wisdom is gained from looking back and using past experience to modify future behaviour.

Most of the top 20 species share similar biological and ecological characteristics, perhaps even common genetic determinates. Yet operational and biological factors, such as high generation turnover, mobility, ability to tolerate plant secondary compounds, high fecundity, short generation times, and genetic

Table 1.3. Top 20 resistant arthropods, ranked by number of unique compounds.

Rank	Species	Family	Order	No. of compounds with reported resistance	No. of references in the APRD	Year of first reported case	Pest of	Common name
1	*Tetranychus urticae*	Tetranychidae	Acari	80	112	1943	Cotton, flowers, fruits, vegetables	Two-spotted spider mite
2	*Plutella xylostella*	Plutellidae	Lepidoptera	76	72	1953	Crucifers	Diamondback moth
3	*Myzus persicae*	Aphididae	Hemiptera	68	78	1955	Fruit, vegetables, trees, grains, tobacco	Green peach aphid
4	*Leptinotarsa decemlineata*	Chrysomelidae	Coleoptera	48	43	1955	Aubergine, pepper, potato, tomato	Colorado potato beetle
5	*Musca domestica*	Muscidae	Diptera	44	35	1947	Urban	Housefly
6	*Boophilus microplus*	Ixodidae	Acari	43	32	1947	Cattle	Southern cattle tick
7	*Blatella germanica*	Blattellidae	Dermaptera	42	65	1956	Urban	German cockroach
8	*Bemisia tabaci*	Aleyrodidae	Homoptera	39	34	1981	Greenhouse, cotton	Whitefly
9	*Panonychus ulmi*	Tetranychidae	Acari	38	68	1951	Fruit trees	European red mite
10	*Aphis gossypii*	Aphididae	Hemiptera	37	24	1965	Cotton, vegetables	Cotton/melon aphid
11	*Culex pipiens pipiens*	Culicidae	Diptera	34	27	1961	Human	Mosquito
12	*Phorodon humuli*	Aphididae	Hemiptera	34	20	1965	Hop, plum	Hop aphid
13	*Helicoverpa armigera*	Noctuidae	Lepidoptera	33	49	1969	Cotton, maize, tomato	Corn earworm
14	*Heliothis virescens*	Noctuidae	Lepidoptera	33	41	1961	Chickpea, maize, cotton, tomato	Tobacco budworm
15	*Culex quinquefasciatus*	Culicidae	Diptera	31	42	1952	Human	Mosquito
16	*Spodoptera littoralis*	Noctuidae	Lepidoptera	30	21	1962	Lucerne, cotton, potato, vegetables	Cotton leafworm
17	*Tribolium castaneum*	Tenebrionidae	Coleoptera	30	31	1962	Stored grain, groundnuts, sorghum	Red flour beetle
18	*Lucilia cuprina*	Calliphoridae	Diptera	25	13	1958	Cattle, sheep	Sheep blowfly
19	*Rhizoglyphus robini*	Acaridae	Acari	22	2	1986	Ornamental plants, stored onions	Bulb mite
20	*Anopheles albimanus*	Culicidae	Diptera	21	12	1964	Human	Malaria mosquito

APRD, Arthropod Pest Resistance Database.

predisposition to resistance coupled with economic factors, such as low economic injury levels, lead to a self-sustaining cycle of high selection pressure and sequential insecticidal applications.

1.7. Resistance by IRAC Mode of Action Group

Another frequently requested output of the APRD is the number of resistance cases classed by IRAC MOA group (Table 1.4). These data are presented herein for the first time and can be used by researchers as well as in a number of

Table 1.4. Number of resistance cases by IRAC mode of action (MOA) classification.

MOA group	No. of cases	% of total	Average no. of cases per annum
1A: Acetylcholinesterase Inhibitors; carbamates	550	7.1	15
1B: Acetylcholinesterase inhibitors; organophosphates	2887	37.3	52
2A: GABA-gated chloride-channel antagonists; cyclodiene organochlorines	1423	18.4	10
2B: GABA-gated chloride channel antagonists; phenylpyrazoles	18	<1	1
3: Sodium-channel modulators; pyrethroids, pyrethrins	1203	15.5	43
3: Sodium-channel modulators; DDT	913	11.8	6
4A: Nicotinic acetylcholine receptor agonists/antagonists; neonicotinoids	57	<1	3
4C: Nicotinic acetylcholine receptor agonists/antagonists; cartap, bensultap	16		<1
5: Nicotinic acetylcholine receptor agonists (not 4); spinosyns	21	<1	1
6: Chloride-channel activators; avermectins, milbemycins	20	<1	<1
7A: Juvenile hormone mimics; juvenile hormone analogues	2	<1	<1
7C: Juvenile hormone mimics; pyriproxyfen	8	<1	<1
8A: Unknown or non-specific – fumigants; alkyl halides	27	<1	0
8B: Unknown or non-specific – fumigants; chloropicrin	55	<1	<1
9A: Unknown or non-specific – selective feeding blockers; cryolite	1	<1	0
9B: Unknown or non-specific – selective feeding blockers; pymetrozine	1	<1	<1
10A: Unknown or non-specific – mite growth inhibitors; clofentezine, hexythiazox	13	<1	<1
10B: Unknown or non-specific – mite growth inhibitors; etoxazole	1	<1	<1
11: Microbial disruptors of insect midgut membranes – includes transgenic crops; *Bacillus thuringiensis* (variety unstated)	69	<1	3
11A1: Microbial disruptors of insect midgut membranes – includes transgenic crops; *Bacillus thuringiensis* var. *israelensis*	4	<1	<1
11A2: Microbial disruptors of insect midgut membranes – includes transgenic crops; *Bacillus sphaericus*	13	<1	<1
11B1: Microbial disruptors of insect midgut membranes – includes transgenic crops; *Bacillus thuringiensis* var. *aizawai*	6	<1	<1
11B2: Microbial disruptors of insect midgut membranes – includes transgenic crops; *Bacillus thuringiensis* var. *kurstaki*	37	<1	1

Continued

Table 1.4. Number of resistance cases by IRAC mode of action (MOA) classification (*continued*).

MOA group	No. of cases	% of total	Average no. of cases per annum
11C: Microbial disruptors of insect midgut membranes – includes transgenic crops; *Bacillus thuringiensis* var. *tenebrionensis*	1	<1	<1
12B: Inhibitors of oxidative phosphorylation, disruptors of ATP formation (inhibitors of ATP synthase); organotin miticides	18	<1	<1
12C: Inhibitors of oxidative phosphorylation, disruptors of ATP formation (inhibitors of ATP synthase); propargite, tetradifon	36	<1	<1
13: Uncouplers of oxidative phosphorylation via disruption of H proton gradient	11	<1	<1
15: Inhibitors of chitin biosynthesis, type 0, Lepidopteran	28	<1	1
16: Inhibitors of chitin biosynthesis, type 1, Homopteran	2	<1	<1
17: Moulting disrupter, Dipteran	14	<1	<1
18A: Ecdysone agonists moulting disruptors; diacylhydrazines	13	<1	<1
19: Octopaminergic agonists	14	<1	<1
20: Coupling site II electron transport inhibitors (Complex III)	1	<1	<1
21: Coupling site I electron transport inhibitors	58	<1	2
22: Voltage-dependent sodium-channel blocker; indoxacarb	12	<1	<1
25: Neoronal inhibitors (unknown mode of action)	1	<1	<1
26: Aconitase inhibitor; fluoroacetate	1	<1	<1
27A: Synergists; P450-dependent monooxygenase inhibitors	10	<1	<1
27B: Synergists; esterase inhibitors (DEF)	5	<1	<1
28: Ryanodine receptor modulator	1	<1	<1
UN[a]: Compounds with unknown mode of action	176	2.3	1
Total	7747	100	

IRAC, Insecticide Resistance Action Committee.
[a]Summation of IRAC MOA classifications UNA, UNB, UNC and UND.

regulatory and market assessment processes by governments and industry. Our data reliably span the years from 1980 to 2007 (Table 1.5). An average of 52 new cases of Group 1B (acetylcholinesterase inhibitors; organophosphates) resistance were reported annually from 1980 to 2007 for the most consistent and frequently reported class of resistance events. Group 3 (sodium-channel modulators) was next in case reports per year with 49, followed by the Group 1A (acetylcholinesterase inhibitors; carbamates) with 15 cases annually. More recently, newer chemistries in Group 4A (nicotinic acetylcholine receptor agonists/antagonists; neonicotinoids) have realized a jump in case reports to almost six per year between 2000 and 2007.

Looking objectively at those chemistries with high or rapidly growing resistance case numbers points to an underlying shift in the pesticides used on highly resistance-prone crops like apples, cotton, and vegetables. Essentially, a comparison based on case events over time can reveal an early susceptibility change associated with the displacement of an older class of chemistries, Group 1, by newer chemistries in Group 4, especially Group 4A. For instance, Group 1 chemistries are being phased out within the USA and also in countries that export commodities into the USA. A major portion of this change was brought on by the passage of the Food Quality Protection Act (1996) in the USA, and the subsequent designation of 'reduced risk' and 'organophosphate replacement' insecticide categories by the USEPA (http://www.epa.gov). When combined, these efforts have accelerated the transition from the Group 1 to Group 4A across both row and specialty crops in the USA. Despite lacking comparable legal incentives, the EU and many other countries are also in a similar transition. Since the data in Table 1.5 are essentially derived by regression analysis of cases (dependent variable) against years (independent variable), they exhibit a lag in communicating the precipitous decline in overall Group 1 use. Thus we would anticipate that reported field cases of Group 1 resistance will actually continue to decline in the future. This may provide new opportunities for after-patent marketers of these older chemistries in situations where they retain registrations and where pest populations that previously exhibited resistance regress due to fitness drag. Despite a decline in the number of cases for Group 1, we will continue to see many cases of resistance for this group, due to the large number of included compounds and the extensive use of Group 1 insecticides.

Responding to earlier trends in insecticide resistance case reports, Mota-Sanchez *et al.* (2002) speculated that Group 4A has the potential to be as resistance prone as Group 1. This prediction was based on the hypothesis that proportionally fewer Group 4A resistance case reports were observed from 1994 to 2000, correlating with the introduction of imidacloprid-based products into several major insecticide markets (Mota-Sanchez *et al.*, 2002) Yet this suggestion may not prove true, with 57 new cases reported since 1994 for Group 4A (imidacloprid, thiamethoxam, acetamiprid, clothianidin, thiacloprid and dinotefuran). These chemistries will certainly be an important group to watch closely, because cross-resistance may also be involved (see Hollingworth and Dong, Chapter 3, this volume) in a number of instances. These case numbers may be especially significant since many of the chemistries in Group 4A have only recently entered large pesticide markets. In the USA for instance, after imidacloprid was introduced in 1994, other chemistries lagged behind: thiamethoxam was released in 2001, acetamiprid in 2002, clothianidin in 2003, thiacloprid in 2003 and dinotefuran in 2004 (Mota-Sanchez *et al.*, 2006; US EPA, 2007a). Given the expansion of many Group 4A chemistries into major markets worldwide and that an average of six resistance cases are currently reported annually from around the world, Group 4A is the class to watch in terms of new resistance case reports in the next 5 to 10 years.

In looking at trends over the past 7 years, Group 3 has also realized a rise in reported cases of resistance to over 42 per year (excluding DDT). In the near future, it will be interesting to see whether the WHO recommendation to begin using DDT again in Africa (Roll Back Malaria Partnership, 2005; Dugger, 2006) in order to accelerate their Roll Back Malaria (RBM) programme has an effect on resistance reports. At this time it is unknown whether this programme and its DDT initiative will result in new cases of mosquito DDT resistance or cross-

Table 1.5. Reported cases of resistance per decade by IRAC mode of action (MOA) classification.

MOA group	Prior to 1980	Prior to 1990	% increase	Prior to 2000	% increase	Prior to 2007	% increase
1A: Acetylcholinesterase inhibitors; carbamates	147	299	103	455	52	550	17
1B: Acetylcholinesterase inhibitors; organophosphates	1442	2132	48	2552	20	2887	11
2A: GABA-gated chloride-channel antagonists; cyclodiene organochlorines	1133	1170	3	1246	6	1423	14
2B: GABA-gated chloride-channel antagonists; phenylpyrazoles, fipronil	0	0	0	2	200	18	800
3: Sodium-channel modulators; pyrethroids, pyrethrins	56	355	534	841	137	1203	43
3: Sodium-channel modulators; DDT	754	896	19	912	2	913	<1
4A: Nicotinic acetylcholine receptor agonists/antagonists; neonicotinoids	0	0	0	17	1700	57	335
4C: Nicotinic acetylcholine receptor agonists/antagonists; bensultap, cartap, nereistoxin analogues	3	5	67	15	200	16	7
5: Nicotinic acetylcholine receptor agonists; spinosyns	0	0	0	1	100	21	2000
6: Chloride-channel activators; avermectins, milbemycins	0	1	100	10	900	20	100
7A: Juvenile hormone mimics; juvenile hormone analogues	0	1	100	2	100	2	0
7C: Juvenile hormone mimics; pyriproxyfen	0	0	0	3	300	8	167
8A: Unknown or non-specific – fumigants; alkyl halides	27	27	0	27	0	27	0
8B: Unknown or non-specific – fumigants; chloropicrin	47	52	11	55	6	55	0
9A: Unknown or non-specific – selective feeding blockers; cryolite	1	1	0	1	0	1	0
9B: Unknown or non-specific – selective feeding blockers; pymetrozine	0	0	0	1	100	1	0
10A: Unknown or non-specific – mite growth inhibitors; clofentezine, hexythiazox	0	0	0	7	700	13	86
10B: Unknown or non-specific – mite growth inhibitors; etoxazole	0	0	0	0	0	1	100
11: Microbial disruptors of insect midgut membranes – includes transgenic crops; *Bacillus thuringiensis* (variety unstated)	2	3	50	49	1533	69	41
11A1: Microbial disruptors of insect midgut membranes – includes transgenic crops; *Bacillus thuringiensis* var. *israelensis*	0	0	0	3	300	4	33
11A2: Microbial disruptors of insect midgut membranes – includes transgenic crops; *Bacillus sphaericus*	0	0	0	7	700	13	86
11B1: Microbial disruptors of insect midgut membranes – includes transgenic crops; *Bacillus thuringiensis* var. *aizawai*	0	0	0	5	500	6	20

Mode of action							
11B2: Microbial disruptors of insect midgut membranes – includes transgenic crops: *Bacillus thuringiensis* var. *kurstaki*	0	2	200	19	850	37	95
11C: Microbial disruptors of insect midgut membranes – includes transgenic crops: *Bacillus thuringiensis* var. *tenebrionensis*	0	0	0	1	100	1	0
12B: Inhibitors of oxidative phosphorylation, disruptors of ATP formation (inhibitors of ATP synthase); organotin midicides (azocyclotin, cyhexatin, fenbutatin oxide)	0	11	1100	15	36	18	20
12C: Inhibitors of oxidative phosphorylation, disruptors of ATP formation (inhibitors of ATP synthase); propargite, tetradifon	30	35	17	36	3	36	0
13: Uncouplers of oxidative phosphorylation via disruption of H proton gradient	6	6	0	7	17	11	36
15: Inhibitors of chitin biosynthesis, type 0, Lepidopteran	0	11	1100	25	56	28	12
16: Inhibitors of biosynthesis, type 1, Homopteran	0	0	0	2	200	2	0
17: Moulting disruptor, Dipteran	0	2	200	10	400	14	40
18A: Ecdysone agonists moulting disruptors; diacylhydrazines	0	2	200	8	300	13	38
19: Octopaminergic agonists	5	10	50	12	20	14	17
20: Mitochondrial complex III electron transport inhibitors	0	0	0	0	0	1	100
21: Mitochondrial complex I electron transport inhibitors	3	3	0	14	367	58	314
22: Voltage-dependent sodium-channel blocker; indoxacarb	0	0	0	1	100	12	1100
25: Neuronal inhibitors (unknown mode of action)	0	0	0	0	0	1	100
26: Aconitase inhibitor; fluoroacetate	1	1	0	1	0	1	0
27A: Synergists; oxidase inhibitors; piperonyl butoxide	[a]					10	
27B: Synergists; esterase inhibitors (DEF)	[a]					5	
28: Ryanodine receptor modulator	0	0	0	1	100	1	0
Undetermined: Compounds with unknown mode of action	153	164	7%	170	4	176	3.50

IRAC, Insecticide Resistance Action Committee.

[a]Decade analysis unavailable.

resistance with synthetic Group 3 chemistries that are being heavily used now. Otherwise, there have been very few new reports of DDT resistance since 1990. Certainly, DDT cross-resistance (see Hollingworth and Dong, Chapter 3, this volume) with other Group 3 chemistries could become a very significant issue in WHO RBM programmes, especially given WHO's continuing efforts to increase the use of bed nets treated with Group 3 chemistries in severe malaria regions in Africa. Rising cases of Group 3 resistance (excluding DDT) could well be in process in the USA as well, as agricultural producers and human and animal protection efforts move towards using Group 3 chemistries, the least expensive alternative to banned Group 1 insecticides.

Group 2A (GABA-gated chloride-channel antagonists) chemistries have obviously also dropped from their position as one of the world's leading resistance-generating insecticide classes, falling from a high of 31% of the world's resistance cases in 1975 to just 18% of the total cases reported by 2007. Almost all of these chemistries from the cyclodiene chemical family were removed from use for various market, residue, and environmental impact reasons. Resistance to cyclodienes will continue to be found owing mainly to just one compound, endosulfan, that is still in use because of its less unfavourable environmental characteristics. In addition, cases of resistance will continue to occur in the chemical family of the phenylpyrazoles (fiproles). Group 1A averaged 15 resistance cases per year followed by ten per year for Group 2A. Certainly Group 1A will continue to fall in field resistance selection episodes as they are phased out of the major markets in the USA, the EU, and their respective trading partners.

These annual resistance case report statistics are informative and give reasonable information following major use and market changes. Thus they can be somewhat predictive of insecticide classes undergoing dramatic use profile changes. With the online, peer-reviewed case publication facilities provided through the APRD (http://www.pesticideresistance.com), perhaps this archiving system will be of some use to resistance managers and policy makers worldwide. Readers may want to review Chapters 3 and 4, which provide a more in-depth review of resistance mechanisms, modelling, and management from both theoretical and practical viewpoints in preparation for these sweeping changes in insecticide use globally.

As in any effort to measure trends in pest management strategies, tactics, and tools where society and legal precedence exist, publication may impact a range of policy and economic features in society. Therefore, interest in, and reporting of, resistance cases has increased within academia, the pesticide industry, pest management associations, government agencies, and retail supply personnel. No doubt the accuracy, utility, and timeliness of resistance information vary with an array of factors, most especially regulatory and trade minimal risk level (MRL) compliance. Yet where there is an absence of IPM, resistance management education efforts, and enumeration statistics, resistance management will suffer. It is no surprise then that more of society is beginning to view resistance as an onerous failure on the part of society to protect valuable susceptible genes in important pest populations. Many pest managers now believe that industry, academia and government share a common mandate to remedy situations where resistance management strategies, tactics and tools are lacking.

Certainly, resistance communication and stewardship of important insecticide molecules is improving from the pesticide industry's perspective, and the development of IRAC by industry is a positive step forward (CropLife International, 2007; IRAC, 2007; IRAC-US, 2007).

1.8. Resistance Reporting by Decade

Another revealing analysis of resistance case reporting in the APRD (http://www.pesticideresistance.com) is the analysis of resistance data by decade (Table 1.5). Obviously, there has been a long history of older insecticide classes like Groups 1A, 2A, 3, UN, 12C, 4B, 4C and 13 that were available prior to 1980. Many of these groups were in use before resistance record keeping was common, and therefore our efforts may yield some spurious enumeration the further back in history we attempt to go. However, based on the efforts of Georghiou and Lagunes-Tejeda (1991) and a very thorough re-review of the older literature by the current authors (which yielded additional citations pre-1980 in the APRD), we are reasonably satisfied with the thoroughness of the pre-1980 data. Obviously the pre-1980 data contain many years for some compounds like DDT, decidedly fewer years for others, and no data for important chemistry groups, like Group 4A, that entered the market in the last decade. In this chapter we do not make an exhaustive effort to sort through all of the old chemistries, but choose to begin a more in-depth analysis beginning arbitrarily in 1980. The data prior to 1980 are presented as a single integer summarizing the resistance cases accumulated up until November 1980. Following this date, we report more accurate and thoroughly reviewed summary resistance enumerations by decade until after 2000, finishing with the 6.5-year period from January 2000 until August 2007.

Obviously, some modes of action representing an array of classes and compounds that have maintained large market presence for decades, like Group 1B, yield extensive case accumulations. Yet we have been able to show a significant downturn in resistance reports for Group 1 through our decade analysis (Table 1.5). Certainly global reports of Group 1A resistance cases are declining and will likely continue to do so into the future. Group 2 resistance reports are declining dramatically in all cyclodiene compounds, except endosulfan for which there have been many cases of resistance reported, mainly from India and China. As previously mentioned, Group 3 chemistries have seen a dramatic increase in reports since 1980, peaking in the decade leading up to 2000, and declining slightly in case numbers since. We expect that resistance case reports will continue at over 50 per year for the next 5 or 10 years. Reports for Group 1A have remained steady since the 1980s. We anticipate some decline in case reports in this group over the next 10 years. Group 11 (microbial disruptors of insect midgut membranes) experienced a peak in the 1990s, but remains the subject of continued scrutiny, especially in genetically modified plants like cotton, maize, and soybeans. Genes for resistance have been detected by an F_2 screen in areas where *Bt* crops have been deployed (Huang *et al.*, 2007). At this moment there is no concern of field failure. However, these findings indicate the potential for field resistance if the proportion with resistance genes overcomes the refuge strategy. We anticipate that these reports will remain constant unless a pest population adapts to a genetically modified plant, which will lead to a very significant jump in case reports. Resistance reports for Group 4 will probably grow, perhaps dramatically, over the next 10 years. Just 6.5 years into the 21st century, this class has seen a jump in resistance case reports annually. Is this an anomaly? Some of the most dramatic examples of field resistance and cross-resistance in this group are found in the Colorado potato beetle (Mota-Sanchez *et al.*, 2006; Alyokhin *et al.*, 2007) and whiteflies (Elbert and Nauen 2000, Nauen *et al.*, 2002; Byrne *et al.*, 2003; Prabhaker *et al.*, 2005), as well as in the laboratory-selected *Nilaparvata lugens* (Liu *et al.*, 2003). Given this class's number of site uses and global sales, resistance cases will continue to rise. Group 12 (inhibitors of oxidative phosphorylation, disruptors of ATP formation (inhibitors of ATP synthase)) resistance cases will continue to be reported, but in low numbers as there are many alternative modes of action in the mite management market. Groups 15 (inhibitors of chitin biosynthesis, type 0, Lepidopteran), 16 (inhibitors of chitin biosynthesis, type 1, Homopteran), 17 (moulting disruptor, Dipteran), and 18 (ecdosyne agonists moulting disruptors; diacylhydrazines) are beginning to see a greater marketplace presence and therefore more resistance reports can be expected. Resistance and cross-resistance issues may rise at a greater rate over the next 10 years, especially as some of these compounds may find uses against Coleoptera, since the insecticide industry seems to have difficulty finding new products effective in their control.

A number of new insecticides with novel modes of action have been marketed (e.g. indoxacarb (Group 22A) and the ryanodine receptor modulators (Group 28)), and many have a relatively narrow spectrum compared with Group 4A, which are effective on a wide spectrum of different arthropod orders. Others, such as Group 10 (unknown or non-specific – mite growth inhibitors), have their principal effect on immature stages. In addition, many of these insecticides are more expensive than conventional insecticides, and they may entail more difficult timing and delivery requirements to be effective. One would expect then that pesticides with these characteristics may not exhibit the same selection pressure as broad-spectrum insecticides, like Group 1B. In fact, the APRD (http://www.pesticideresistance.com) does record fewer overall reports of resistance for such compounds. However, this relaxed resistance setting may be changing as the APRD data suggest a tendency towards an increased rate of resistance cases for some of the newer, relatively specialized, chemistries.

We have summarized the resistance cases for each IRAC MOA group in a linear format to facilitate reader reference (Table 1.4). From this perspective, it is rather apparent that Groups 1, 2, 3, 4, 5, 6, 7 and 8 constitute more than 95% of the resistance cases ever reported. Several summary deductions seem obvious. First, given six decades, billions of tonnes of insecticides and miticides applied, many thousands of square miles treated, and the relatively narrow modes of action utilized, it is a wonder that microevolution was not more robust! Secondly, given the relatively narrow range of modes of action deployed and those that have generated most of the world's resistance cases, there is good reason to believe that future resistance management through the deployment of a greater number of modes of action may yield greater stability. Perhaps, 60 years from now, microevolutionary adaptation in pest populations will be an old problem, because the pest management community has progressively accepted resistance management tactics and tools as a routine element of IPM programmes. Finally, our agriculture, human and animal protection, and other diverse pesticide treatment arenas and endeavours have a wealth of resistance experience to draw from. How well the pest management community learns their lessons may well depend on our responses to these resistance data.

1.9. Resistance Management and Future Enumeration

Some commodities, like the US potato industry and their pesticide suppliers, have taken significant strides to educate and encourage their constituents to use sound resistance management strategies, tactics, and tools in preserving important compounds for control of their key pests (NPC, 2007). Targeting available resources from both private and public sources could result in more and earlier reports of resistance, leading to the rapid adoption of resistance avoidance strategies (Roush, 1989; Fitt, 2003). It is our hope that the APRD and associated newsletter could help amend the headlong evolution of pest adaptations to insecticides and miticides.

Perhaps an extreme illustration of the social, economic, political, and regulatory factors involved in developing resistance management procedures and recommendations can be gleaned from US and EU efforts to secure registrations of the first genetically engineered plant varieties incorporating elements of *Bt* insecticidal toxin proteins (see Andow *et al.*, Chapter 5, this volume). This process, a more or less cooperative effort by industry and the interested regulatory, academic, and environmental organizations, spurred the development of a series of working resistance management strategies in many countries. Ironically and unfortunately, this period of intense resistance management interest was narrowly focused upon genetically engineered plants alone, excluding the application of these same principles to conventional insecticides. Thus, an opportunity to extend comprehensive resistance management programmes to a range of insecticide tools was lost. Nevertheless, the outcome of the genetically modified plant deliberations led US regulators to install mandatory resistance reporting and management plans for all genetically engineered plants. Their aim was to assure the environmental community, growers, regulators, and industry that these steps would ameliorate resistance (US EPA, 2007b). Ironically, it was the US organic and environmental communities, seeking to preserve the effectiveness of *Bt*, that provided some of the political will and media attention that eventually led to the regulatory reforms centred on protecting *Bt* for future generations of organic growers (see Thompson *et al.*, Chapter 6, this volume).

Quite apart from genetically engineered plants, regulatory authorities directly and indirectly influence resistance development by increasing or decreasing the chain of arthropod exposures, whereby a pest population can be repeatedly selected over and over again from site to site, field to field, or crop to crop. Selection pressure can mount with increased spatial and temporal mar-

keting and management practice. For instance, the US EPA 'use site' classification (introduced in the 1996 Federal Insecticide, Fungicide, Rodenticide Act)
is designed to define a unique active ingredient's registered use on a particular
crop, animal, stored product, right of way, etc. In other words, 'use site' is a classification of an application site whereby a pesticide can be legally used according to its label. Most regulatory authorities throughout the world have similar
structures to organize their regulatory burden and harvest fees. In the USA, the
more USEPA registered 'use sites' on an insecticide's label, the greater the potential exposure and selection pressure a chemistry may induce on one or more
arthropod species. Thus 'area-wide' and 'sequential seasonal' exposure analysis may be helpful in targeting situations and pesticide use sequences where
arthropods, like those in our top 20 list (Table 1.3), are placed under severe selection pressure. In these instances, arthropods may essentially be forced to
adapt or become extinct in an area or a cropping system (Forrester *et al.*, 1993).

In response to such situations, cotton, melon and vegetable pest managers
in Australia (Roush, 1989; Fitt, 2003) and then those in Israel and Arizona,
USA pioneered spatial, temporal, and human infrastructures to avert resistance
development to critical crop management tools that are effective against a range
of pests (Dennehy and Denholm, 1998). In these intensive-production regions,
a wise insecticide deployment strategy can protect the crops involved, valuable
pest management tools, and the susceptibility genes in targeted pest populations. Alternatively, a complacent industry and society has often allowed producers to create resistance rapidly, thereby denying a long use-life of key products
both in their own and other's production commodities. This situation is especially important where producers are attempting to avoid disease transmission
through insect vectors. The success of these cited resistance prevention efforts
may be a harbinger of future similar investments in many cropping and human
health management systems. Therefore, regulatory agencies, the pest management and supply industries, consultants, environmental activists, producer organizations and academics alike can significantly influence resistance enumeration
and management.

Perhaps one could assert that effective management of resistance is more
about common sense, social infrastructures that allow resistance management
implementation, empowering local community leadership and effective communication around a common purpose where many elements of business, production and society join in to achieve a worthwhile goal. When resistance is
assumed and resources are mobilized together with a relatively simple and
straightforward resistance management plan, like diversifying mortality factors
in target pest populations and avoiding sequential selection episodes, growers
or pest managers can achieve sustainable, economical, and effective resistance
management. Recording resistance cases and the information that surrounds
these events allows forensic backtracking and case history development, which
can be used to help educate generations of pest managers about resistance
(NPC, 2007) and provide information to pest managers on how to avoid such
episodes in the future. Historical and current resistance information, together
with spatial and timely resistance information, can play a role in such resistance
management systems; and the APRD may have a unique place in the future,

particularly in resistance education and development of the incentives for sound resistance management programmes by providing a ready information source that can inform those interested in the history, resistance potential, and methods that have been used in the past to overcome resistance development.

Acknowledgements

The APRD and *Resistant Pest Management Newsletter* would not have been possible without collaboration and support from members of the IRAC over years of development. IRAC committee leaders like Paul Leonard and Gary Thompson have been excellent colleagues in these endeavours, contributing their time, energy, and insights into the APRD development processes.

The editors would also like to thank the USDA Cooperative State Research, Education, and Extension Service, specifically Michael Fitzner and Rick Meyer, for support of the concept and limited long-term funding for the development of the APRD.

The editors and authors would also like to express sincere appreciation to Ashley Wight for her excellent editorial assistance.

Pat Bills was a tireless early worker on the Database. Without his seminal ideas and expertise, the APRD would never have been cast in electronic media. We also appreciate Qang Xue's input in the transition of the Database to a web-accessible, case-submission system. Lee Duynslager, an information specialist at Michigan State University, provided excellent technical assistance in this process and Lee is a key member of our Executive Team.

We are especially indebted to many students over the years who contributed untold hours in the processes of analysing, extracting, and posting over one million data into various electronic file formats in the APRD: Lisa K. Losievsky, Saunte T. Sutton, Lauren E. Keinath, Mareck Ulicny and Robbie Wang. In the same manner, we would also like to acknowledge all of the Resistant Pest Management Newsletter editors, Erin Vidmar, Theresa A. Baker, Jeanette E. Wilson and Abbra A. Puvalowski, who assisted in the publication of over 600 articles, abstracts, and other communications informing the world of arthropod resistance development and management.

References

African Insect Science for Food and Health (2007) Human Health. http://www.icipe.org/human_health/index.html (accessed December 2006).

Alyokhin, A., Dively, G., Patterson, M., Castaldo, C., Rogers, C., Mahoney, M. and Wollam, J. (2007) Resistance and cross-resistance to imidacloprid and thiamethoxam in the Colorado potato beetle *Leptinotarsa decemlineata*. *Pest Management Science* 63, 32–41.

Brown, A.W.A. (1958) *Insecticide Resistance*. World Health Organization, Geneva, Switzerland.

Busvine, J.R. (1968) Resistance to organophosphorous insecticides in insects. *International Symposium Phytopharm Phytiat,* pp.605–620.

Byrne, F.J., Castle, S., Prabhaker, N. and Toscano, N.C. (2003) Biochemical study of resistance to imidacloprid in B biotype *Bemisia tabaci* from Guatemala. *Pest Management Science* 58, 868–875.

Croft, B. (1990) *Arthropod Biological Control Agents and Pesticides.* Wiley, New York.

CropLife International (2007) Resistance management. http://www.croplife.org/issue.aspx?issue=0b6532cb-0eff-45d5-8190-757d5720ab1b&activity=09927e10-d23f-4fc4-99c1-12998754de05&wt.cg_n=Stewardship&wt.cg_s=Stewardship%20of%20crop%20protection%20products&wt.ti=Resistance%20management (accessed August 2007).

Crow, J.F. (1960) Genetics of insecticide resistance: general considerations. *Miscellaneous Publications of the Entomology Society of America* 2, 69–74.

Dennehy, T.J. and Denholm, I. (1998) Goals, achievements and future challenges of the Arizona Whitefly Resistance Management Program. In: *Proceedings of the 1998 Beltwide Cotton Production Research Conference.* National Cotton Council, Memphis, Tennessee, pp. 68–72.

Downs, A.M.R., Stafford, K.A., Ravenscroft, J.C. and Coles, G.C. (2002) Widespread insecticide resistance in head lice to the over-the-counter pediculocides in England, and the emergence of carbryl resistance. *British Journal of Dermatology* 146, 88–93.

Dugger, C.W. (2006) WHO supports wider use of DDT versus malaria. *The New York Times*, 16 September.

Elbert, A. and Nauen, R. (2000) Resistance of *Bemisia tabaci* (Homoptera: Aleyrodidae) to insecticides in southern Spain with special reference to neonicotinoids. *Pest Management Science* 56, 60–64.

FAO (1980) *Recommended Methods for Measurement of Pest Resistance to Pesticides.* FAO Plant Production and Protection Paper No. 21. Food and Agriculture Organization of the United Nations, Rome.

Ffrench-Constant, R.H. and Roush, R.T. (1990) Resistance detection and documentation: the relative roles of pesticidal and biochemical assays. In: *Pesticide Resistance in Arthropods.* Chapman and Hall, New York, pp. 4–38.

Fitt, G.P. (2003) Deployment and impact of transgenic *Bt* cotton in Australia. In: Kalaitzandonakes, N.G. (ed.) *The Economic and Environmental Impacts of Agbiotech: A Global Perspective.* Kluwer, New York, pp. 141–164.

Forrester, N.W., Cahill, M., Bird, L.J. and Layland, J.K. (1993) Management of pyrethroid and endosulfan resistance in *Helicoverpa armigera* (Lepidoptera: Noctuidae) in Australia. *Bulletin of Entomological Research* Suppl. 1, R1–R132.

Georghiou, G.P. and Lagunes-Tejeda, A. (1991) *The Occurrence of Resistance to Pesticides in Arthropods. An Index of Cases Reported Through 1989.* Food and Agriculture Organization of the United Nations, Rome.

Huang, F.N., Leonard, B.R. and Andow, D. (2007) Sugarcane borer (Lepidoptera: Crambidae) resistance to transgenic *Bacillus thuringiensis* maize. *Journal of Economic Entomology* 100, 164–171.

IRAC (2007) Introduction to IRAC. http://www.irac-online.org/IRAC/Structure.asp (accessed August 2007).

IRAC-US (2007) Homepage on Internet. http://www.irac-online.org/IRAC_US/Home.asp (accessed August 2007).

Liu, Z.W., Han, Z.J., Wang, Y.C., Zhang, L.C., Zhang, H.W. and Liu, C.J. (2003) Selection for imidacloprid resistance in *Nilapavarta lugens*: cross-resistance patterns and possible mechanisms. *Pest Management Science* 59, 1355–1359.

Melander, A.L. (1914) Can insects become resistant to sprays? *Journal of Economic Entomology* 7, 167–173.

Mota-Sanchez, D., Bills, P.S. and Whalon, M.E. (2002) Arthropod resistance to pesticides: status and overview. In: Wheeler, W.B. (ed.) *Pesticides in Agriculture and the Environment*. Marcel Dekker, Inc., New York, pp. 241–272.

Mota-Sanchez, D., Hollingworth, R.M., Grafius, E.J. and Moyer, D. (2006) Resistance and cross-resistance to neonicotinoid and spinosad insecticides in the Colorado potato beetle, *Leptinotarsa decemlineata* (Say) (Coleptera: Chrysomelidae). *Pest Management Science* 62, 30–37.

Nauen, R., Stumpf, N. and Elbert, A. (2002) Toxicological and mechanistic studies on neonicotinoid cross resistance in Q-type *Bemisia tabaci* (Hemiptera: Alerodidae). *Pest Management Science* 58, 868–875.

NPC (2007) Environmental Programs: Pesticide Environmental Stewardship Program. http://www.nationalpotatocouncil.org/NPC/programs_environmental.cfm (accessed November 2007).

Pimentel, D., McLaughlin, L., Zepp, A., Lakitan, B., Kraus, T., Kleinman, P., Vancini, F., Roach, W.J., Graap, E., Keeton, W.S. and Selig, G. (1991) Environmental and economic impacts of reducing US agricultural pesticide use. In: Pimentel, D. (ed.) *Handbook of Pest Management in Agriculture*. CRC Press, Boca Raton, Florida, pp. 679–718.

Pimentel, D., Acquay, H., Biltonen, M., Rice, P., Silva, M., Nelson, J., Lipner, V., Giordano, S., Horowitz, A. and D'Amore, M. (1993) Assessment of environmental and economic impacts of pesticide use. In: *The Pesticide Question: Environment, Economics, and Ethics*. Chapman & Hall, Inc., New York, pp. 47–84.

Prabhaker, N., Castle, S., Henneberry, T.J. and Toscano, N.C. (2005) Assessment of cross-resistance potential to neonicotinoid insecticides in *Bemisia tabaci* (Hemiptera: Aleyrodidae). *Bulletin of Entomological Research* 95, 535–543.

Roll Back Malaria Partnership (2005) *World Malaria Report 2005*. World Health Organization/United Nations Children's Fund, Geneva, Switzerland.

Roush, R.T. (1989) Designing resistance management programs: how can you choose? *Pesticide Science* 26, 423–441.

Sawicki, R. (1987) Definition, detection and documentation of insecticide resistance. In: Ford, M.G., Holloman, D.W., Khambay, B.P.S. and Sawicki, R.M. (eds) *Combating Resistance to Xenobiotics: Biological and Chemical Approaches*. Ellis Horwood Ltd, Chichester, UK, pp. 105–117.

Theiling, K.M. and Croft, B.A. (1988) Pesticide side-effects on arthropod natural enemies: a database summary. *Agriculture, Ecosystems & Environment* 21, 191–218.

Tomlin, C.D.S. (1997) *The Pesticide Manual*. British Crop Protection Council, Farnham, UK.

Townsend, A.M. (2001) Network cities and the global structure of the Internet. *American Behavioral Scientist* 44, 1697–1716.

US EPA (2007a) Pesticides: Topical and Chemical Fact Sheets. http://www.epa.gov/pesticides/factsheets/index.htm (accessed August 2007).

US EPA (2007b) Introduction to Biotechnology Regulation for Pesticides. http://www.epa.gov/oppbppd1/biopesticides/regtools/biotech-reg-prod.htm (accessed November 2007).

Whalon, M.E. and McGaughey, W.H. (1998) *Bacillus thuringiensis*: use and resistance management. In: Ishaaya, I. and Degheele, D. (eds) *Insecticides with Novel Modes of Action: Mechanism and Application*. Springer, Berlin, pp. 106–137.

WHO (1957) *Expert Committee on Malaria, Seventh Report*. WHO Technical Report Series No. 125. World Health Organization, Geneva, Switzerland.

2 Documentation of Pesticide Resistance in Arthropods

D. Mota-Sanchez, M.E. Whalon, R.M. Hollingworth
and Q. Xue

Department of Entomology, Michigan State University, East Lansing, Michigan, USA

2.1 The Arthropod Pesticide Resistance Database (APRD)

Resistance is a widespread phenomenon whereby arthropod populations develop the ability to avoid the lethal effects of normally fatal concentrations of one or more pesticides. The occurrence of pesticide resistance frequently leads to the increased use, overuse, and even misuse of pesticides that pose a risk to the environment, phytosanitation, market access, global trade, and public health. It can also result in serious economic loss and even social disruption. A 1984 study initiated by the National Research Council's Board on Agriculture made 16 recommendations, one of which identified the worldwide need for a central and permanent 'data bank' of resistance information (NAP, 1986). To accomplish this objective a resistance database was developed (see Chapter 1, this volume, for the history and development of the Arthropod Pesticide Resistance Database (APRD)). In August 2006, the old Michigan State University (MSU) database was transferred to a new system (http://www.pesticideresistance.com) designed by the authors of this chapter. The new data system includes features to meet the needs of comprehensively reporting pesticide resistance worldwide. In general, data are collected to determine which pests have developed resistance, the pesticide or control tactic, cross- and multi-resistance linkages, the means of resistance detection, and a scaling factor of these data in time and space. A design process for the database (Fig. 2.1) was drafted that dictates its exact contents and functions. The underlying design specification includes the ability to generate analysis and summaries of pest resistance data for publication and distribution. One key addition was made in the data entry fields, allowing the pesticide mode of action to be recorded, based on the categories of the Insecticide Resistance Action Committee (IRAC) (IRAC, 2007). Another feature is a system allowing the web-based input of cases of resistance by registered database users which enables resistance reporting in close to real time. The development of the database in its current form was supported through a partnership between IRAC, the US Department of Agriculture (USDA) Cooperative State Research, Education, and Extension Service's Integrated Pest Management Programme and MSU.

By including as many data as possible, we intend to maximize the utility for referencing emergent cases of resistance. The APRD contains the following key pieces of information.

1. The arthropod (class, order, family, species, common name).
2. Active ingredient and mode of action.
3. Origin of the resistant population (if the population was resistant when collected in the field, if the population was created solely by selection and/or genetic manipulation in the laboratory, or if the population was selected further in the laboratory after collection).
4. If known, the mechanism of resistance (insensitive target site, metabolism, decreased penetration, behavioural, unknown or other).
5. The location (country, state or province, county, prefecture or nearest city, site, description of the area, postal or telephone area code, or GPS coordinates).
6. Collection site, host, and date of collection.
7. Bioassay method, date of bioassay, life stage tested, and sex.
8. Values of LC_{50}, LD_{50}, LT_{50}, LC_{95}, LD_{95}, LT_{95}, confidence intervals, resistance ratios and discriminating doses.
9. Impact of the resistance (none, low, medium, high, severe, or unknown).
10. Reference to the document reporting the case.
11. Cross-resistance to other compounds.

What constitutes a 'case' of resistance in the database?

How 'resistance case' is defined is a critical point in documenting and recording resistance. The authors of this chapter define a resistance case as the identified resistance of a single population of any life stage of a single species to a single active ingredient at a given time and location. There are critical points and combinations of events that can complicate this definition. Resistance is a population genetic phenomenon that is variable in both time and space. Any designation of a report of resistance as a discrete case involves a series of judgements, frequently based on incomplete evidence, and may be rather arbitrary in the face of a lack of clear definition of the criteria we use in a published report. The following sections consider the most common challenges in making this assignment. Where there is doubt, the approach currently is to designate a report as a case in order to be comprehensive in collecting and cataloguing cases.

Resistance ratio, discriminating doses and biochemical tests
The strength of the APRD relies upon the expertise of the reviewers of manuscripts and the editorial boards of publications, as well as upon our own review of the median lethal dose (LD_{50}), the median lethal concentration (LC_{50}), the

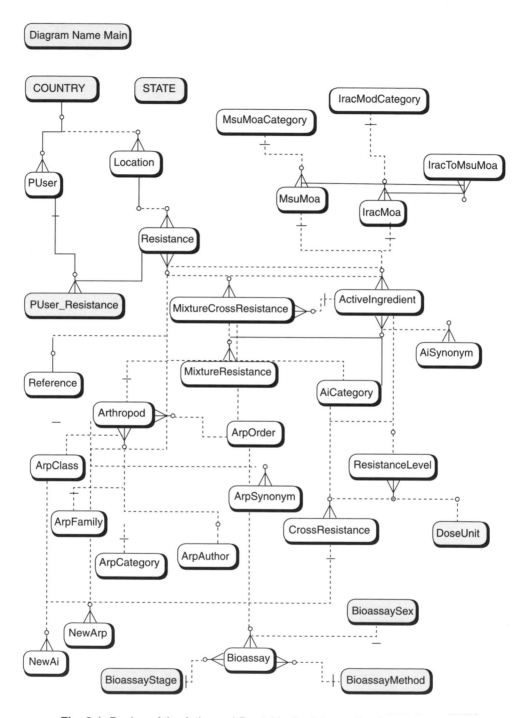

Fig. 2.1. Design of the Arthropod Pesticide Resistance Database.

median lethal time (LT_{50}), or their associated values at the 95% response level (LD_{95}, LC_{95}, LT_{95}). Data from discriminating doses are also considered. Methodological and statistical review is our primary method for determining whether a proposed incidence of resistance is valid. This review involves examining the statistical differences between resistant and susceptible reference colonies for previously unreported species, compounds, and/or geographical sites. A commonly reported measure of resistance is the resistance ratio (RR), which is the ratio of the toxic dose against the tested strain (defined by the statistic used, e.g. LD_{50}, LD_{95}, LC_{50}, KD_{50} or LT_{50}) to that of a known susceptible strain. We have used a RR of 10 or greater as a general threshold for declaring a 'case' of resistance. This ratio is arbitrary but reasonably conservative, because ratios tenfold or larger do not necessarily impact pest control in the field (see review of Ffrench-Constant and Roush, 1990). However, in some cases we also include reports with RR smaller than 10 when the authors were clear that this was high enough to cause significant problems of control in the field. In any case, we have included a new feature in the database that includes the resistance ratios and the impact of the resistance. Thus in the future we will be able to associate more ratios of resistance with the impact of resistance in the field.

Frequently scientists define resistance by using discriminating doses (e.g. LD_{90}, LC_{90}, LD_{99} or three times the LD_{90} of the susceptible strain, among others). The ability of a population to survive such doses that would normally kill a high proportion of the susceptible strain indicates the presence of resistance. A population which survives a pesticide treatment at three times the LD_{99} has greater resistance than a population which survives the LD_{90} of a susceptible population. We have a link in our new system that categorizes the impact of resistance identified by discriminating dose methods, and a case is considered valid where the authors consider that resistance will have an impact on field control and particularly where the discriminating dose test is used as a field diagnostic test to monitor resistance. Biochemical tests to detect the presence of higher enzymatic activity, or receptor insensitivity, or DNA tests to differentiate resistant from susceptible individuals may also be considered to identify a case of resistance where the findings are positive and the trait being assayed clearly relates to resistance *in vivo* in that population.

Geographic considerations: location/population

Another critical factor in defining a case of resistance is related to the concept of population, since the scientists who publish resistance papers frequently refer to the pest 'population' as the group of insects collected in a specific site. Futuyma (2005) defined population as:

a group of conspecific individuals that occupy a more or less well defined geographic region and exhibit reproductive continuity from generation to generation; ecological and reproductive interactions are more frequent among these individuals than with members of other populations of the same species.

However, in resistance studies, the spatial definitions are often not clear, or the limits are overlapping. In addition, to meet the requirements to unambiguously define a population would be very expensive because of the amount of studies. With the potential limitations of fulfilling the Futuyma definition, the APRD and its editorial board recognize a 'case of resistance' as the reduced sensitivity to a particular insecticide from insects sampled from a limited geographical area. Even at closely located sites, resistance reports may be counted as separate cases if there is evidence for a low probability of gene flow between these sites, as for example with Colorado potato beetles in a series of isolated potato fields or houseflies in poultry houses. We have also expanded our definition of a resistance case to subspecies level. For instance, cases of resistance of subspecies of mosquitoes, *Culex pipiens*, are reported in the database because the experts in the area classified them as such.

Time considerations: repeated detection in the same population

This is important because the dynamics of the resistance is important in the field failure of the compound. Sometimes low levels of resistance do not result in field failure, but serve however as a warning of a potential problem in the future. We have registered cases of initial low levels of resistance in a single location. With time and increasing insecticide selection pressure, the levels of resistance increased causing the failure of the insecticide. Since the dynamics of the resistance is critical in the resistance management we have also included cases of resistance through time for a single population, of any stage, of a single species, in a single site, to a unique pesticide. Repeated reports of resistance to the same insecticide in the same insect from the same location are quite uncommon in the database, but these would not now be considered separate cases. However a change in the qualitative nature of the resistance, such as the expansion of cross-resistance to a new group of compounds in this location, would be counted as a new case.

Multi-resistance/cross-resistance

We have also considered cases of resistance where the resistance to one compound also confers cross-resistance to other compounds. For instance, in Long Island, New York, Colorado potato beetles developed resistance to imidacloprid and cross-resistance to seven other neonicotinoid insecticides despite these compounds never having been used against Colorado potato beetle (Mota-Sanchez *et al.*, 2006). Since the definition of a case is resistance to a single compound in a single species at a unique location, each of the seven bioassay results would be counted as a new case.

Laboratory versus field selection: are they all 'cases'?

In order to be comprehensive, we have included all cases of resistance whether developed in the field or in the laboratory. The total number of reports involving selection solely in the laboratory is relatively low (perhaps 5% of the total). However with some compounds such as *B. thuringiensis* toxins, the percentage of reports from laboratory selection is much higher, and the combined total will obviously overestimate the cases of resistance occurring by selection in the field. As now constituted, the ARPD has the capability to count these two routes to resistance independently if desired.

Are we underestimating the magnitude of global arthropod pesticide resistance?

We have recorded 7747 cases of resistance that correspond to 1694 references (Fig. 2.2) spanning 398 journals, reports, proceedings and dissertations. We have made considerable effort to include all reported cases of resistance. Articles reviewed were principally in English, but we have reviewed some in Spanish and French. However, there are sure to be cases that are not included in our database because of reports in other languages such as Chinese, Russian, and Arabic, among others. In addition, reports of resistance are frequently published only as local reports from experimental stations in developing countries where the diffusion of information is limited to the local area or region. Even more important, many cases of resistance probably go unstudied and unreported. Finally, cross-resistance is usually not fully explored in resistance studies, so this also underestimates cases by the criteria of the database. For these reasons, it is likely that the total number of cases in the database represents only the tip of the iceberg and significantly underestimates the true number.

As we said, there are a potential number of cases reported in other languages or published only in local reports. However, the main cause of the underestimation of resistance prior to the development and implementation of the APRD was the software limitations of our predecessor and the previous database. For example, prior compilations of resistance cases did not document the

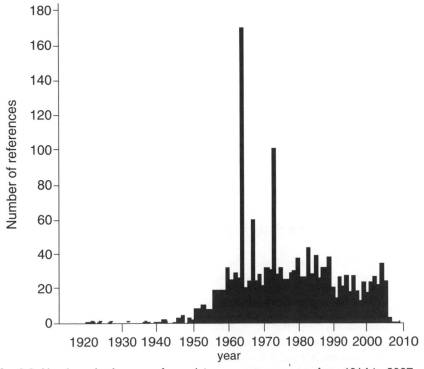

Fig. 2.2. Number of references for resistance cases per year from 1914 to 2007.

geographic location of the case in as fine detail as is now achievable with the APRD. Most of the cases were cited to a level of country and state or province, but finer resolutions were not registered before December 2006. Now our system allows us to include the GPS coordinates from which populations are sampled. Given these circumstances and a sampling of articles, we estimate that the real number of cases in the literature could increase from 7747 total cases of resistance to at least 15,000 to 30,000 cases. As documentation of geographic location becomes more precise, these cases will be entered into the database.

Other important factors that may impact the enumeration of resistance cases are the following:

1. Resistance reports may be biased to choose the 'hot spots' where the worst-case scenario for developing resistance occurs. While this is often true, particularly for new compounds, the motivation for scientists to report resistance in other situations is reduced once this type of worst-case resistance is detected.
2. Economic and educational problems in developing countries dramatically reduce the monitoring of resistance despite the impact of the resistance on the farmers in the form of insecticide field failure.

It is important to mention that the focus of monitoring resistance has been in arthropods that are pests of agriculture, forestry, and veterinarian and human health. Despite the efforts to record resistance cases, few studies focus on predators, parasitoids, pollinators, and other non-target arthropods. An exception is the many studies of *Drosophilla melanogaster* as a model for resistance development. Use of insecticides would select non-target organisms along with target pests. Given the tremendous complexity of arthropod diversity in crops from both temperate and tropical latitudes, it is a difficult task to estimate the number of other arthropods potentially being selected. Non-target organisms such as predators, parasitoids, and pollinators are not the main focus of the efforts to monitor resistance. However, the existence and movement of those arthropod species in an insecticide-treated environment may act as an important driving force for microevolution and ecological interactions. Given these additional situations, one could speculate that the total cases of resistance perhaps would increase significantly to six figures.

2.2 Web-based Case Submission

The ARPD also includes a web-based case entry system (http://www.pesticideresistance.com) that is open to anyone who wishes to record one or more cases of resistance. This entry system consists of a secure password-gated system, an automated scientific and web-based editorial process, and a process that assigns a unique accession number for each case accepted into the database. The case submission process is divided into simple and clear steps. Users can track their submissions and a case can be cloned to avoid filling forms repeatedly with similar information. Case reviewing is done completely through the online system. E-mails are sent automatically to the editors to inform them of the cases to be reviewed. Approved cases are entered into the public database for searching and reporting. The cases entered by a user are sent to the APRD where a web page displays the entire database content. In this way we hope to make the database more comprehensive and also to capture the advance of resistance in something approaching real time.

The APRD will have a committee led by resistance specialists that will initially be generated from the *Resistance Pest Management Newsletter* subscribers who will review submissions. Up to now, we have received communications from scientists from public and private industry all over the world who are willing to participate in the editorial process to review cases of resistance. In addition, we will look for experts in the USDA, universities, and private sectors.

2.3 Summary

With a growing human population that will require more food, pressure to produce alternatives to petroleum-based fuels, and limited resources in new areas open to agriculture, there will inevitably be more pressure to increase the yields in existing production areas. However, arthropods and other organisms will continue to compete with man for these resources. Application of insecticides and other tools, including genetically modified organisms, has been deployed to control these pests. However, since herbivorous arthropods have been evolving for millions of years to defeat toxins, it is not a surprise to have 553 species and 7747 cases of resistance, most of which have been recorded over the last 60 years of intensive pesticide use. It is our intention that this effort in reporting arthropod pesticide resistance should contribute in designing better alternatives for resistance pest management; and in the end contribute to the world's effort to reduce hunger, and improve human and animal health and food security. Therefore, we invite all the community related to agriculture or animal and human health to use our database and submit cases of resistance to it.

References

Ffrench-Constant, R.H. and Roush, R.T. (1990) Resistance detection and documentation: the relative roles of pesticidal and biochemical assays. In: *Pesticide Resistance in Arthropods*. Chapman and Hall, New York, pp. 4–38.

Futuyma, D.J. (2005) *Evolution*. Sinauer Associates, Sunderland, Massachusetts.

IRAC (2007) Mode of Action Classification Scheme. http://www.irac-online.org/Crop_Protection/MoA.asp (accessed November 2007).

Mota-Sanchez, D., Hollingworth, R.M., Grafius, E.J. and Moyer, D. (2006) Resistance and cross-resistance to neonicotinoid and spinosad insecticides in the Colorado potato beetle, *Leptinotarsa decemlineata* (Say) (Coleoptera: Chrysomelidae). *Pest Management Science* 62, 30–37.

NAP (1986) *Pesticide Resistance: Strategies and Tactics for Management*. National Academy Press, Washington, DC.

3 The Biochemical and Molecular Genetic Basis of Resistance to Pesticides in Arthropods

R.M. HOLLINGWORTH AND K. DONG

Department of Entomology, Michigan State University, East Lansing, Michigan, USA

3.1. Introduction

In the last three decades remarkable strides have been made in understanding pesticide resistance in arthropods, first at the level of biochemical and physiological mechanisms, and more recently at the level of molecular genetics and genomics. The explosion of molecular genetic studies has revealed many details of the mechanisms of resistance, both at the individual and population levels, and has provided new insights on the microevolutionary processes that have produced them; yet in many cases it has also revealed unexpected complexities that are proving challenging to unravel.

Studies over several decades have demonstrated that there are a limited number of mechanisms by which resistance is achieved and that these are often monogenic. These mechanisms primarily involve either the increased metabolic destruction of the toxic agent or changes in its site of action that decrease or eliminate the target site's responsiveness to attack. A number of subsidiary mechanisms have also been described which include decreased rates of uptake, enhanced internal binding of the active insecticide to sites without toxic effects, and resistance to the deleterious effects following the interaction of the toxicant with its site of action. The advent of the genomics revolution in some cases has allowed the precise identification of the genetic and molecular changes that have created a resistant strain as well as providing new tools to monitor these traits within populations. However, a second outcome of genomic analysis has been to reveal a previously unexpected complexity in the array of potential isoforms of receptors, ion channels and enzymes that are the keys to the toxicity of and resistance to insecticides. Some current challenges are therefore to understand the significance of this abundance of isoforms in the physiology of the insect. For example, although we can identify a mutation in an acetylcholine receptor subunit that decreases the binding of insecticidal agonists and is associated with resistance to them, we know very little about the subunit composition of any native receptor in insects, where such receptors are located within the nervous system or their specific function there, or how different isoforms are adapted to different functions and vary in expression within the life of the insect. In addition, much of this fund of knowledge has been gained with just a handful of insect species where genomics analysis is most advanced, particularly *Drosophila* and *Anopheles*. These are both dipterans and neither are insects of agricultural importance. Although it is reasonable to suppose that other insects are similar in general, enough differences exist even within the insects with annotated genomes that caution is necessary in extrapolating too aggressively from a very limited number of species. By comparison, much less is known about the genomics of such important agricultural pests as mites, lepidopterans and coleopterans. Finally, we have progressed much more slowly in understanding the total pharmacokinetics of insecticide action and how multiple resistance mechanisms interact to produce a given level of resistance.

The topic of this chapter has been the subject of several multi-authored books in the last decade (Denholm *et al.*, 1999; Ishaaya, 2001; Clark and Yamaguchi, 2002). The remarkable recent advances in the understanding of the molecular mechanisms and genomics of insecticide resistance have been reviewed by Feyereisen (1995), Soderlund (1997), Oakeshott *et al.* (2003), and Ffrench-Constant *et al.* (2004). Many individual aspects of the biochemistry and genetics of resistance are included in a recent multi-volume compendium on insect molecular science (Gilbert *et al.*, 2005). Resistance in specific insects that have been the major subjects of genomic studies is reviewed by Wilson (2001) for *Drosophila*, by Hemingway *et al.* (2004) for mosquitoes, and, more broadly, for insect vectors of human disease by Hemingway and Ranson (2000). The present review focuses primarily on advances that have occurred in the 10 years up to mid-2006.

3.2. Changes in Insecticide Sites of Action

With the probable exception of mitochondrial uncouplers, every potent insecticide has one or more specific binding sites on a critical macromolecule within the insect. The disruption of function in this macromolecular site of action that results from insecticide binding initiates the cascade of events that leads eventually to the death of the insect. Changes in the ability of the insecticide to bind to the site of action, or to affect its functions after binding, must lead to significant changes in the overall impact on the insect. There is ample evidence that such changes are a primary cause of resistance to many types of insecticide. Almost invariably, those so far established have taken the form of point mutations leading to changes in critical amino acid residues in the receptor molecule rather than to changes in the level of expression of existing receptors. Where the site is critical to the normal functioning and survival of the insect, it seems that a limited number of mutations may be possible that maintain physiological competence while decreasing sensitivity to the insecticidal ligand. However, there are few cases where it appears that a functional target site is not critical for the survival of the insect even though its interaction with the insecticide leads to mortality. In these cases, mutations with a more radical change in the activity of the site, either through mutation or decreased expression, will not be disadvantageous, and even the complete elimination of the gene product (null mutation) is a viable pathway to high-level resistance. For example, see Gahan *et al.* (2001) demonstrating the loss of a cadhedrin-binding protein for *Bacillus thuringiensis* toxin in *Heliothis virescens*; Orr *et al.* (2006), who describe the loss of the *Dm 6* nicotinic acetylcholine receptor subunit that binds the spinosyns in resistant *Drosophila*; and the loss of a binding protein for juvenile hormone analogues in *Drosophila* (Wilson and Ashok, 1998).

Acetylcholinesterases

The role of acetylcholinesterase (AChE) as the primary mechanism for removing the excitatory neurotransmitter, acetylcholine (ACh), from cholinergic synapses and its role as the target site for organophosphate and carbamate inhibitors in both vertebrates and insects have long been established (Giacobini, 2000). The accumulation of ACh that results from the inhibition of AChE leads to the prolonged stimulation and, in many cases, the desensitization of the ACh receptors (AChR), eventually to severe neurological disruption, and ultimately to death. The homodimeric molecular architecture of the AChE enzyme and its catalytic site were first described based on crystallographic analysis in *Torpedo* (Sussman *et al.*, 1991), and parallel key structural features in the *Drosophila* enzyme were described by Harel *et al.* (2000) and reviewed by Oakeshott *et al.* (2005). ACh and, presumably, insecticidal inhibitors approach the active site through a narrow cleft or gorge that is hydrophobic in nature and leads to a series of residues that bind ACh and carry out its hydrolysis to acetate and choline with extremely high efficiency. Key amino acids of the active site, which are highly conserved, include a tryptophan residue (and probably other aromatic amino acids) that interacts with the trimethylammonium cationic group of ACh in what was once termed 'the anionic site', although it now appears that aromatic (π-electron) rather than electrostatic forces predominate in binding this quaternary ammonium group of ACh. A 'catalytic triad' of residues is responsible for the hydrolysis of the substrate, consisting of a serine residue, which is acetylated by ACh, and a histidine which enhances the nucleophilicity of the serine hydroxyl in combination with a glutamate residue. The reaction is completed by the facilitated hydrolysis of the acetyl group from the serine by an activated water molecule. An oxyanion hole accommodates the carbonyl group of ACh and helps to stabilize the tetrahedral transition states of acetylation and deacetylation through hydrogen bonding.

Organophosphates (OPs) and carbamates are simply poor substrates for this enzyme with very low turnover numbers. They readily phosphorylate or carbamylate the critical serine residue, but hydrolysis to remove the phosporyl or carbamyl moiety and release the active enzyme is very slow, with a half-life of an hour to days. The active site of the enzyme is therefore occupied and incapable of hydrolysing its normal substrate. Several allosteric sites also exist on

AChE that modify the activity of the enzyme but these are not generally thought to be the primary site of action of insecticides.

Higher Diptera have only one gene for AChE (termed *ACE*) whereas other insects, including mosquitoes and mites, have two, although probably only one (*ACE-1*) is expressed in, and is active in removing ACh from, the central nervous system (Russell *et al.*, 2004). Surprisingly, this is not the orthologue of the single enzyme in the higher Diptera, so it appears that two quite distinct genes encode for the critical neuronal AChE even in different members of the same order, Diptera (Weill *et al.*, 2002). There is a recent report of a third *ACE* gene in a tick (Fournier, 2005). The functions of the additional *ACE* genes are unknown but they do not appear to be associated with resistance.

A decrease in the sensitivity of AChE to inhibition by some OPs and carbamates leading to resistance has been demonstrated in a broad range of insects and mites (Russell *et al.*, 2004; Fournier, 2005; Oakeshott *et al.*, 2005). Examples of the up-regulation of AChE are known and may play a modest role in resistance (see e.g. Charpentier and Fournier, 2001), but high-level effects are due to point mutations that decrease inhibitor efficiency. These resistance-related mutations occur at multiple sites in and around the catalytic zone. A total of 14 point mutations so far discovered in insect *ACE* genes that provide some degree of insecticide resistance are listed by Fournier (2005). In some cases the level of resistance is quite low. These mutations are discussed in detail with consideration of their possible mechanistic significance by Oakeshott *et al.* (2005). Generally they involve steric effects that may hinder inhibitor access to, or binding at, the active site. These mutations vary considerably in the level of resistance they confer and in the spectrum of effects among different insecticidal inhibitors. It is not uncommon for the same mutation to cause increased resistance to some inhibitors and enhanced sensitivity to others. For example, using the sequence numeration for the mature *Torpedo californica* enzyme, the Phe115Ser mutation in the *Drosophila* enzyme provides a resistance factor of 30 for paraoxon, but leads to a threefold increase in sensitivity to dichlorvos, and the mutation Gly368Ala in *Drosophila* provides tenfold resistance to chlorpyriphos-oxon, but has no effect on sensitivity to dichlorvos, and causes a fourfold increase in sensitivity to coumaphos-oxon. Likewise, a Phe→Tyr mutation at position 330 in the *Drosophila* sequence provides strong (90-fold) resistance to diazoxon but a threefold increased sensitivity to omethoate (Menozzi *et al.*, 2004; Fournier, 2005). It is not uncommon for multiple mutations to exist in the same resistant enzyme sequence. This combination of mutations generally results in an increase in the level of resistance (Menozzi *et al.*, 2004; Fournier, 2005). Not only does this intensify resistance to individual compounds but it also tends to remove the negatively correlated cross-resistance to other inhibitors that may be seen with single mutations.

Russell *et al.* (2004) have postulated that these cases of resistance can be classified into two major groups: Pattern I, leading to resistance primarily to carbamates; and Pattern II, leading to resistance about equally to both OPs and carbamates, although some OP-specific resistance can also occur. In a few cases of each pattern, gene sequencing has identified the molecular nature of the alteration leading to the lowered sensitivity to inhibitors. Although it is not possible yet to relate with full confidence the mechanism by which these structural changes alter sensitivity to inhibitors, Pattern I mutations may involve changes in the active site, such as a common Gly→Ser mutation in the oxyanion hole, whereas Pattern II changes may result in a constriction of the cleft leading to an active site that limits the access of inhibitors and, presumably, of ACh itself. It is also interesting to speculate that the loss of one of the two insect *ACE* genes (*ACE-1* homologue) and apparent transfer of neural function to the *ACE-2* homologue in higher *Diptera* has changed the potential for mutational resistance development within this group.

Insensitivity to inhibitors may be accompanied by a reduced ability to hydrolyse ACh. Whether this is always deleterious to the organism is unclear since it is generally considered that, as in vertebrates, AChE is present in insects at a level considerably in excess of that needed for basic neurological functions under normal physiological conditions. However, in several cases, significant fitness costs have been described (Oakeshott *et al.*, 2005). In the peach-potato aphid, pleiotropic effects of altered AChE have been reported, including decreased reproduction and a heightened response to the aphid alarm pheromone (Foster *et al.*, 2003). In the case of *Culex pipiens*, the typical mutation in *ACE-1* converts a glycine residue to serine close to the catalytic site (Weill *et al.*, 2003). This has a clear fitness cost including prolonged developmental time, morphological changes, and decreased survival over winter (Bourguet *et al.*, 2004). Perhaps as a response, duplication of the *ACE-1* locus has occurred in some resistant forms of *Culex* so that they contain both the wild-type and mutant forms. This probably provides higher AChE activity in the absence of selection and yet leaves an inhibitor-resistant form that ensures survival during insecticide exposure (Lenormand *et al.*, 1998). A second example of genetic modification to restore fitness comes from a study of the effect of the four most common resistance-related *ACE* mutations in *Drosophila melanogaster* alone and in combination (Shi *et al.*, 2004). The effects of these mutations on substrate hydrolysis were variable. However, in every case, the stability of the mutant AChE was lower than that of the wild type. The combination of decreased activity and decreased stability would probably be particularly deleterious to the nervous system. However, in some cases, subsequent mutations partially restored fitness by reversing the loss of enzyme catalytic activity. Finally, the overproduction of a mutated AChE in *Drosophila* also leads to a restoration of the total ACh hydrolytic capacity lost by an initial resistance-producing mutation (Charpentier and Fournier, 2001).

Nicotinic acetylcholine receptors

As with vertebrates, two broad types of receptors for the neurotransmitter ACh exist in insects: muscarinic receptors that are linked to slower G-protein-mediated postsynaptic actions, and nicotinic receptors that open ion channels through the neuronal membrane leading to a rapid but transient shift in membrane polarization. Unlike the situation in vertebrates, these receptors are known to exist solely in the central nervous systems of insects. Only nicotinic receptors (nAChRs) are believed to be the direct target for current insecticides. Three groups of insecticides impact nAChRs, each of which acts at a different site on the receptor ion channel complex. These compounds are the neonicotinoids, with structures analogous to that of nicotine itself (Jaschke and Nauen, 2005); the spinosyns, which are macrocyclic lactones acting allosterically to activate a subgroup of nAChRs at a site distinct from that of the neonicotinoids (Salgado and Sparks, 2005; Orr *et al.*, 2006); and a group of compounds derived from the marine annelid toxin, nereistoxin, such as cartap, that blocks these receptors non-competitively (Lee *et al.*, 2003).

In vertebrates, and probably in insects also, the nAChRs are homo- or hetero-pentameric membrane-spanning units with a central ion channel. At least two ACh-binding sites are present, formed by adjacent subunits, and both must be occupied to initiate channel opening (Karlin, 2002). There is an abundance of potential subunits: in some insects there are at least ten nAChR subunit genes with the added potential for alternative splicing variants and RNA-editing of the transcripts. These, in theory, could be combined in a very large number of different pentameric combinations, although the subunit composition is not known for any native receptor (Sattelle *et al.*, 2005). Difficulties in obtaining functional expression of these receptors in cell systems using only insect subunits have seriously hampered investigation of their functional properties. Although the co-expression of insect and vertebrate subunits can generate functional receptors, it is not yet clear that these chimeric receptors have the same properties as all-insect combinations.

Resistance to the nereistoxin groups is known, but detailed analysis of the mechanism is not available. Resistance to the spinosyns occurs in a number of insects. In resistant strains of *D. melanogaster* produced by mutagenesis, resistance is attributable to several different mutations in a gene encoding the Dα6 nAChR subunit, in some cases leading to the complete loss of this site of action for the insecticide (Orr *et al.*, 2006). Resistance to neonicotinoids is increasing with their widespread use and has been reviewed by Nauen and Denholm (2005). In some cases, increased degradative metabolism, particularly by monooxygenases, is involved. However, in other cases, evidence for an alteration in the sensitivity of the nervous system to the effects of the insecticide has been reported (Liu *et al.*, 2005, Mota-Sanchez *et al.*, 2006). In the brown planthopper, resistance to imidacloprid has been attributed to a mutation in AChR subunits that decreases specific binding to nervous system membrane preparations (Liu *et al.*, 2005). The occurrence of a single mutation (Tyr151Ser) in a conserved region believed to be involved in ACh binding was found in two subunits, Nlα1 and Nlα3, and correlated well with the presence of imidacloprid resistance. When these insect α-subunits were co-expressed with rat β2-subunits to form chimeric receptors, a virtually complete loss of imidacloprid binding was observed compared with the same subunits from susceptible insects. Why both subunits should show the same mutation and whether both must be mutated to generate resistance are significant questions from the viewpoint of the probability of developing site-of-action resistance activation. Although this is the first demonstration of a resistance-related mutation in an nAChR causing neonicotinoid resistance, it is unlikely to be the last.

GABA receptors and other ligand-gated chloride channels

γ-Aminobutyric acid (GABA) is the primary inhibitory transmitter in the insect nervous system, with GABA-ergic transmission occurring both within the central nervous system and at the neuromuscular junction. These rapid-acting GABA receptors (GABARs) are linked to channels that gate chloride ions and thus they cause hyperpolarization upon opening. A second group of GABARs in insects gate cations and are excitatory in effect, but little is known of their functions or their role in insecticide action, if any. Their pharmacology is distinct from that of the GABARs gating chloride channels (Gisselmann *et al.*, 2004). The structures, functions, molecular genetics and interactions of inhibitory GABARs with insecticides have been reviewed by Buckingham and Sattelle (2005) and Buckingham *et al.* (2005).

The inhibitory GABARs are pentamers with the subunits arranged to form a central ion-conducting pore. Each subunit probably has four transmembrane domains. The receptor has several distinct binding sites for xenobiotics. Multiple forms of the subunits exist *in vivo* although relatively little is known of the number, nature, specific functions, or localization of the native receptors. Studies with vertebrate subunits expressed in different combinations in *Xenopus* oocytes indicate that the properties of the receptors, including their pharmacology and response to insecticides, are strongly dependent on the subunit composition (see e.g. Ratra *et al.*, 2002). The number of subunit genes is not yet clear because it is difficult to identify them based solely on sequence data, but there are several reasonable possibilities (Buckingham *et al.*, 2005). Even though the number of subunit genes may be low compared with vertebrates, as with other ionotropic receptors, a considerable range of subunit structures may exist because of the possibility of alternative splicing and RNA-editing (Buckingham *et al.*, 2005).

Other ionotropic receptors that gate chloride ions also exist in the insect nervous system with either glutamate or histamine as their activator. In particular, glutamate-gated channels (glutamate H receptor) (GluHR) have an important role in insecticide action. For some compounds, such as the avermectins, actions on both GluHR and GABAR are probably involved in their overall toxicity to insects. To confuse this situation further, there is evidence for the *in vivo* co-assembly of Rdl, a GABAR subunit, with a GluHR subunit in *D. melanogaster* (Ludmerer *et al.*, 2002) to form an apparently mixed-type receptor. The physiological functions and significance of such hybrid receptors in the action of, and resistance to, insecticides are not yet known.

Several groups of insecticides act, either primarily or importantly, at the GABAR ion channel complex. The chlorinated cyclodienes typified by dieldrin, chlordane and endosulfan, lindane and the more recent phenylpyrazole (fiprole) family, act as non-competitive receptor antagonists, binding at a site near the ion channel and acting to stabilize an inactive state of the channel. The resulting inhibition of the GABA-ergic transmission leads to strong excitation marked by hyperactivity and convulsions. The variable, but generally low, degree of cross-resistance between dieldrin and fipronil (Bloomquist, 2001), and the different nature of their competition for binding with other convulsant ligands, indicate that the binding sites for dieldrin and fipronil are not identical although they may have elements in common. Several macrocyclic lactones, including avermectin derivatives (Rugg *et al.*, 2005) and the spinosyns (Salgado and Sparks, 2005), enhance GABA-ergic effects by binding to other allosteric sites on the GABAR. The resulting increase in inhibitory inputs causes decreased activity and prolonged paralysis. These lactones may also have other critical sites of action on GluHRs or nAChRs. Evidence has also been presented that avermectins have significant actions on histamine-gated chloride channels in insects (Iovchev *et al.*, 2002). The physiological and biochemical actions of all these compounds have been reviewed by Bloomquist (2001, 2003).

The biochemical and genetic basis for resistance to dieldrin and related compounds has been investigated in *Drosophila* in an exemplary series of studies by Ffrench-Constant and co-workers (reviewed in Ffrench-Constant *et al.*, 2000). The gene involved in resistance was isolated from a field-collected population of *D. melanogaster*. This gene, termed *Rdl*, is widely expressed in the central nervous system. Interestingly, and revealing that our understanding of GABARs in insects is still quite fragmentary, *Rdl* is not expressed in muscle. Muscle must contain other, undescribed, subunits that confer sensitivity to cyclodienes since they have a powerful blocking action on GABAR at the neuro-muscular junction (Buckingham and Sattelle, 2005). On the other hand, since mutations in *Rdl* yield high resistance to cyclodienes and fipronil, the neuro-muscular receptors can hardly be a critical site of action for these compounds. *Rdl* encodes a GABAR subunit, and orthologues are found very widely among other insect species. Alternative splicing sites at two locations in the *Drosophila* gene could yield four different gene products. When expressed in oocytes or insect cell lines, these subunits create functional channels that respond to GABA and gate chloride ions. The native form of the ion channel is blocked by micromolar concentrations of cyclodienes and fiproles. In resistant *Drosophila*, a point mutation in *Rdl* converts an alanine residue to serine in the pore-forming region of the channel located on the M2 domain. This reduces the binding potency of insecticidal antagonists and also decreases the rate of receptor desensitization (Hosie *et al.*, 1995). Homologous mutations are seen in many other insects and this also correlates with dieldrin and fipronil resistance. Although this point mutation seems to be easily the most common cause of resistance to these convulsant insecticides, an alanine to glycine shift at the same locus has been reported in the aphid, *Myzus persicae* (Anthony *et al.*, 1998), and recently in a laboratory-selected strain of *D. simulans* (Le Goff *et al.*, 2005). In addition, a new mutation locus was discovered in the *D. simulans* strain involving a threonine to methionine conversion in the M3 domain. With oocyte-expressed receptors, both mutations gave substantial resistance to fipronil although their effects in relation to repeated GABA application were different and complex. Very high levels of resistance to fipronil occurred when both mutations were present together. The homomeric receptor composed of five Rdl units can be expressed in oocytes and insect cell lines. The resulting complex appears to have many of the properties expected for native receptors and it seems clear that the mutation of this subunit is a major factor in resistance to GABAR-blocking insecticides. However, native receptors are likely to be more complex in their subunit structure (Buckingham *et al.*, 2005) and much is yet to be learnt of their properties and specific role in insects and insecticide action and resistance.

Fipronil is also a potent blocker of some non-desensitizing glutamate-gated chloride channels in the cockroach thoracic ganglion neurons, whereas dieldrin has no effect on them (Ikeda *et al.*, 2003; Zhao *et al.*, 2004). This may help explain the lower toxicity of fipronil to vertebrates that have inhibitory GABARs but not inhibitory glutamate receptors (GluRs), although differences in the relative binding potency of fipronil to Rdl, and in degradative metabolism, may also favour toxicity to insects (Buckingham and Sattelle, 2005). Also, the fact that Rdl-based cross-resistance between dieldrin and fipronil is not well correlated could be explained if fipronil has an additional major toxic effect on these GluRs that dieldrin lacks.

Resistance to the macrocyclic lactones, such as abamectin, occurs independently of that to the convulsant GABAR-active insecticides, and cross-resistance to fipronil or dieldrin is not typically seen. Moreover, these compounds also have a major action on GluHR rather than GABAR or on native heterologous combinations of both GluR and GABAR subunits (Rugg *et al.*, 2005). Although resistance to avermectin analogues has been reported in several insects and mites in the field, the mechanisms and cross-resistance patterns are varied and not always well defined (Clark *et al.*, 1995; Rugg *et al.*, 2005). Metabolic mechanisms seem the most common basis for resistance, and no well-established example of a changed binding site has been reported so far, although point mutations in GluR subunits in ivermectin-resistant nematodes have been reported to reduce the binding of ivermectin (Dent *et al.*, 2000; Njue *et al.*, 2004). In *Drosophila* selected in the laboratory, a modest level of resistance to ivermectin is associated with a proline to serine mutation in the structural gene for the GluHR subunit, GluClα, and with a decreased affinity of ivermectin to binding sites in head membranes, and a decreased sensitivity of these subunits to ivermectin when expressed in *Xenopus* oocytes (Kane *et al.*, 2000).

Voltage-dependent sodium channels

The movement of sodium ions across the axonal membrane is a key factor in the development of a nerve action potential. The opening of sodium-specific voltage-dependent channels occurs as the wave of depolarization induced by an approaching action potential reaches a critical value. After opening, the sodium channels are rapidly inactivated, which cuts off the inward flow of sodium current and thus limits the depolarization. Three groups of insecticides impact this process by their actions on voltage-dependent sodium channels (VDSCs). Despite their quite different structures, DDT and its relatives and the pyrethroids have a common mode of action. They modify sodium channels mainly by slowing channel deactivation (i.e. the closure of activation gate) and trap sodium channels in the open configuration, resulting in prolonged channel opening evidenced by a large tail current associated with repolarization (Vijverberg *et al.*, 1982; Narahashi, 1988). This may lead to the repetitive discharge of action potentials, or, if depolarization is high enough, to a complete block of axonal transmission (Narahashi, 2002). Either action has highly disruptive effects on the nervous system.

Indoxacarb is an oxadiazine insecticide that impacts VDSCs by a different mechanism. Indoxacarb and its *N*-decarbomethoxylated metabolite, DCJW, block sodium channels of various insect and mammalian neurons by stabilizing VDSCs in an inactivated (closed) state (for references see Wing *et al.*, 2005). Studies using the rat $Na_v1.4$ and cockroach sodium channel variants expressed in *Xenopus* oocytes confirmed the state-dependent block of both mammalian and insect sodium channels by DCJW (Silver and Soderlund, 2005; Song *et al.*, 2006). These studies suggest that the actions of indoxacarb and DCJW share some similarities with those of local anaesthetics, which inhibit sodium currents by binding preferentially to the inactivated state of the sodium channel (Hille, 1977). No cases of resistance to indoxacarb based on changes in target site have yet been reported, but mutations in the VDSC are a common cause of resistance to DDT and the pyrethroids (Soderlund *et al.*, 2002; Soderlund and Knipple, 2003; Khambay and Jewess, 2005; Soderlund, 2005; Dong, 2007).

One major mechanism of pyrethroid resistance, known as knockdown resistance (kdr), reduces neuronal sensitivity to DDT and pyrethroids and confers insect cross-resistance to all pyrethroids. Since its first discovery in a housefly strain in 1951, the kdr-type mechanism has been reported in many medically and agriculturally important insect pests (Soderlund and Bloomquist, 1990). The most significant earlier research on the kdr mechanism revealed nerve preparations of kdr insects to be less sensitive to pyrethroids and also less sensitive to veratridine and aconitine, two site 2 sodium channel neurotoxins (see references in Soderlund and Bloomquist, 1990). The major advance in kdr research in the last decade started with several groups independently demonstrating a tight linkage between pyrethroid resistance and *para*-orthologous sodium channel genes in several insect species (Taylor *et al.*, 1993; Williamson *et al.*, 1993; Dong and Scott, 1994; Knipple *et al.*, 1994). The *para* gene was isolated from *D. melanogaster* mutants exhibiting a temperature-sensitive paralytic phenotype (Loughney *et al.*, 1989) and encodes a voltage-gated sodium channel in *Xenopus* oocytes (Feng *et al.*, 1995; Warmke *et al.*, 1997). Like their mammalian counterparts, insect sodium channels contain four repeated homologous domains (I–IV), each having six membrane-spanning segments (S1–6) connected by intracellular or extracellular loops.

Nucleotide and deduced amino acid sequences of full-length or partial cDNAs of the sodium channel genes from various arthropod species were determined to reveal any amino acid changes that are associated with the kdr phenotype. The first kdr mutation from leucine (L) to phenylalanine (F) in domain II segment 6 (IIS6) was identified in the housefly (L1014F) and the German cockroach (L993F). This mutation was later detected in other insect pest species, but not in arachnid species, varroa mite, or southern cattle tick. Mutations from L to histidine (H) or serine (S) at the same position were found in *H. virescens* and *C. pipiens*. In addition to the L1014F mutation, highly pyrethroid-resistant houseflies (the super-kdr flies) carry a second mutation (methionine (M) to threonine (T) at 918) in the linker region connecting IIS4 and IIS5 (Williamson *et* al., 1996). The same mutation, M to T, along with the first kdr mutation, was later detected in pyrethroid-resistant horn flies, *Haematobia irritans* (Guerrero et al., 1997). These early findings raised the possibility that one or two common mutations are responsible for pyrethroid resistance in all insects. However, more new sodium channel mutations have since been found to be associated with pyrethroid resistance. So far, a total of 26 pyrethroid resistance-associated point mutations have been identified in the sodium channel genes from more than a dozen arthropod species (Soderlund, 2005). It appears that both common and unique sodium channel mutations are associated with pyrethroid resistance in different insect and arachnid species. Most of the mutations were found either in transmembrane segments IS6, IIS5, IIS6, and IIIS6, or intracellular loops or linkers close to the transmembrane segments. So far, 12 of these pyrethroid resistance point mutations have been confirmed to be involved in reducing the sensitivity of insect sodium channels to pyrethroids using the *Xenopus* oocyte expression system (Dong, 2007). The level of reduction in pyrethroid sensitivity by these kdr mutations appears to correlate with the level of kdr resistance *in vivo*. For example, the first kdr mutation, L1014F (housefly) and L993F (cockroach), in IIS6 reduced the sensitivity of housefly, cockroach, and *Drosophila* sodium channels by five- to tenfold, whereas the double mutations, L1014F and M918T, in the super-kdr trait, reduced the channel sensitivity to pyrethroids by 100-fold (Lee et al., 1999; Vais et al., 2000). Two additional cockroach mu tations, E434K (glutamate to lysine at 434) and C764R (cysteine to arginine at 764), along with the first kdr L993F mutation, were found in highly resistant German cockroach populations. The triple-mutation cockroach sodium channel was almost insensitive to pyrethroids (Tan et al., 2002). Most strikingly, the F to isoleucine (I) mutation in IIIS6 (F1519I), first identified in highly pyrethroid-resistant southern cattle ticks, completely abolished the cockroach sodium channel sensitivity to pyrethroids (Tan et al., 2005).

How do the kdr mutations alter sodium channel sensitivity to pyrethroids? These mutations could confer sodium channel resistance to pyrethroids by reducing the binding of pyrethroids to the sodium channel and/or by counteracting the action of pyrethroids via a binding-independent mechanism (e.g. by altering sodium channel gating). Naumann (1998) proposed a 'horseshoe model' for pyrethroid action, based on pyrethroid structure–activity relations, stereochemical information, and toxicity data. This model predicted that the binding site (pocket) for active pyrethroids, in which the pyrethroid curves into a horseshoe shape, is possibly at an aromatic residue of the sodium channel backbone, and suggested a common active conformation for the structurally diverse pyrethroids at the sodium channel. Although a high-affinity pyrethroid-binding site was detected on the α-subunit of rat brain sodium channel preparations 10 years ago (Trainer et al., 1997), direct measurement of pyrethroid-binding affinity and capacity to insect sodium channels has not been achieved, because the high lipophilicity of pyrethroids results in extremely high levels of non-specific binding to membranes and filters and consequently masks any specific pyrethroid binding (Pauron et al., 1989; Rossignol, 1988; Dong, 1993). Vais et al. (2000, 2001) showed that the kdr (L1014F) in IIS6 and the super-kdr mutations M918T in the IIS4–S5 linker and T929I in IIS5 all enhanced closed-state inactivation, and thereby reduced channel opening, which is required for the action of pyrethroids. These mutations also reduced the binding affinity for open channels (Vais *et al.*, 2000, 2003). A recent study (Tan *et al.*, 2005) provides direct evidence for the involvement of F1519 in IIIS6 and L993 in IIS6 in forming part of the elusive pyrethroid-binding site. Utilizing the competitive binding of active (1*R-cis*) and inactive (1*S-cis*) permethrin isomers to the sodium channel, Tan *et al.* (2005) showed that L993F in IIS6 and F1519W reduced pyrethroid binding to the cockroach sodium channel, and that an aromatic residue at position 1519 is required for the action of pyrethroids. These results support this long-standing model and suggest that F1519 is probably the aromatic residue in Naumann's horseshoe model. A recent study used the crystal structures of potassium channels to generate a housefly sodium channel model in the open conformation (O'Reilly *et al.*, 2006). This model highlights the role of IIS4–S5 linker and the IIS5 and IIIS6 helices in pyrethroid binding, supporting the involvement of M918, T929 and F1519 in pyrethroid binding.

Previous electrophysiological studies (Lund and Narahashi, 1982; Vais *et al.*, 2000) suggested that pyrethroids interact with the sodium channel at multiple sites. Lund and Narahashi (1982) examined competitive and non-competitive interactions of *cis*- and *trans*-tetramethrin (a type I pyrethroid) with sodium channels in squid axons. Their analysis suggested that there are separate *cis* and *trans* pyrethroid-binding sites on the sodium channel (Lund and Narahashi, 1982). It was also hypothesized that the super-kdr mutation (M918T) of the housefly sodium channel eliminated one of the pyrethroid-binding sites (Vais *et al.*, 2000). However, results from Tan *et al.* (2005) are most compatible with a single pyrethroid-binding site with multiple binding steps. First, F1519I completely abolished the channel sensitivity to all eight structurally diverse pyrethroids examined (Tan *et al.*, 2005). Therefore, binding at F1519 appears to be a prerequisite for pyrethroid binding and action on other sites. A lack of the initial binding of pyrethroids to F1519 seems to prevent subsequent pyrethroid interactions with other residues, such as L993 and M918. Secondly, the F1519I mutation not only altered the channel sensitivity to 1*R-cis*-permethrin, but also reduced the channel sensitivity to 1*R-trans*-permethrin (unpublished results), indicating that F1519 is critical for both *cis*- and *trans*-pyrethroids. A similar multi-step binding model has been proposed for the interaction between noradrenaline and β2 adrenergic receptors (Kobilka, 2004; Liapakis *et al.*, 2004).

Juvenile hormone receptors

The juvenile hormone analogues (juvenoids) are a class of compounds that mimic the actions of juvenile hormones (JHs) in the insect and thus initiate or continue JH-like actions when they are no longer physiologically appropriate for the development of the insect. This disrupts larva–pupa and pupa–adult moults, as well as affecting many other processes such as embryonic development, caste determination, reproductive success and behaviour in specific insects (Dhadialla *et al.*, 1998, 2005; Goodman and Granger, 2005). Despite significant effort, the unequivocal identification of JH receptors remains elusive (Flatt et al., 2005; Goodman and Granger, 2005).

Resistance to juvenoids is not very common, particularly under field conditions, although it has been reported in houseflies, flour beetles, whiteflies and mosquitoes, and is reviewed by Mohandass et al. (2006). Although the mechanism is often unclear, it appears that the juvenoid resistance is often a result of cross-resistance related to changes in the metabolism and perhaps rates of uptake of the insecticide. However, Cornel et al. (2002) reported a very high level of resistance to methoprene in wild populations of mosquitoes in central California after 20 years of use. The resistance was not reversed by typical synergists that inhibit esterases and P450 oxygenases, so it is possible that an altered site of action could be involved.

Resistance to these compounds has also been developed in D. melanogaster by mutagenesis. Flies having one of several variants of the Met (methoprene tolerance) trait are highly resistant to several juvenoids, and to JH itself, with no cross-resistance to other insecticides. Resistance is related to a decrease in the binding affinity of JH III to a protein in the cytosol of the fat body (Shemshedini and Wilson, 1990). In resistant insects there is a reduction in the transcript from the Met gene that codes for this protein, which has the characteristics expected of a transcription factor of the PAS family. In one mutagenized strain, the resistant larvae have a null mutant of Met, in which the Met transcript is entirely lacking (Wilson and Ashok, 1998). The flies are still viable in this strain although reductions in oocyte development, fecundity, and competitiveness have been reported. This indicates that the Met protein is not critical for survival and development even though it is a major insecticidal target. Recently it has been shown that JHs (particularly JH III) and methoprene bind directly and with high affinity to the Met gene product and activate it when expressed in insect S2, cells where it locates in the nucleus (Miura et al., 2005). Whether the Met product represents a functional JH receptor or acts as a transcription factor in the transduction of JH signalling is currently uncertain (Flatt et al., 2005). Wilson et al. (2006) recently reported that an unexpectedly large variety of scattered point mutations can be produced in the Drosophila Met gene that lead to resistance to methoprene, and forecast that this could lead to the rapid development of resistance if these compounds are more widely deployed.

Insecticidal microbial toxins

Several types of proteinaceous bacterial toxins are currently used for insect control. The two principal sources are Bacillus thuringiensis (Bt) and B. sphaericus (Bs). Bt strains produce a large variety of crystal protein δ-endotoxins that are used either in conventional spray-on applications or are genetically engineered into plants. Over 200 different toxin-producing Cry genes have been identified. The variety, nomenclature and uses of Bt toxins have been reviewed by Bravo et al. (2005). Several different groups of endotoxins are produced which tend to be highly specific for different orders of insects such as lepidopterans, coleopterans, and mosquitoes.

After solubilization and proteolytic activation in the insect gut, it is believed that *Bt* endotoxins exert their toxicity by binding to receptors on the midgut epithelial cells of sensitive insects. The aggregation of several of these toxin molecules then leads to their insertion into the luminal membrane to form a pore. The resulting loss of ionic control causes osmotic disruption of the midgut cells leading to swelling and lysis, which is lethal. Several types of receptor for the *Bt* toxins toxic to lepidopterans have been identified on the midgut epithelial cell surface, including glycosylphophatidylinositol-anchored aminopeptidase-N (APN), a digestive enzyme, and cadherins, known to act as intercellular adhesion molecules. The process of creating the pore after receptor binding has occurred is complex and not fully understood, but may involve cadhedrin and APN receptors operating in sequence. The cadherins are probably the initial *Bt* monomer-binding site leading to further specific proteolysis of the monomers that is critical for their aggregation into tetramers. Binding of the tetramers to APN may then help to insert them into the membrane and lead to pore formation (Bravo *et al.*, 2005). Multiple receptor types are clearly involved in the actions of different *Bt* toxins since binding site-based resistance to Cry1A toxins does not lead to cross-resistance to Cry1C toxins (Ferré and Van Rie, 2002).

The various mechanisms of resistance to *Bt* so far discovered have been reviewed by Ferré and Van Rie (2002), Bravo *et al.* (2005), and Griffitts and Aroian (2005). Only one insect species, the diamondback moth (*Plutella xylostella*), is known to have become resistant to *Bt* toxins through field exposure and the detailed mechanism for this resistance is not known. However, it is monogenic and partially recessive and is typically characterized by a dramatic decrease in the specific binding of radiolabelled Cry1A toxins to brush border membrane vesicles isolated from the resistant midguts (Ferré and Van Rie, 2002; Sayyed *et al.*, 2004, 2005). Many examples of *Bt* resistance have been developed by laboratory selection and many of the mutations involved carry a considerable fitness cost (e.g. Oppert *et al.*, 1997; Ferré and Van Rie, 2002), which would be strongly unfavourable in field populations. Although laboratory selection sheds light on the potential mechanisms for resistance that might occur in the field and provides evidence for the mechanism of action of toxins, it is by no means certain that these same alterations would be selected and survive under field conditions. For example, the genetic basis of resistance developed in diamondback moth populations in the field is not identical to that developed in laboratory-selected lepidopterans, even though both may involve reduced receptor binding (Baxter *et al.*, 2005).

The main mechanism for resistance selected in the laboratory is a reduction in the binding affinity between the toxin and its receptor (Ferré and Van Rie, 2002). The correlation of resistance patterns to changes in binding characteristics is made more complex by the presence of multiple binding sites in the same insect for distinct but related toxins; for example, at least four binding sites have been identified in *P. xylostella* for different Cry1 proteins. Recent studies have identified the cause of some of these changes in binding affinity to be mutations in cadherin genes. The genomic changes identified include the introduction of additional sequences into the coding region through retrotranposon action leading to gene knockout associated with extremely high resistance to Cry1Ac (Gahan *et al.*, 2001), and point mutations in a cadherin gene in *H. virescens* (Xie *et al.*, 2005) leading to the modification of the Cry1Ac toxin-binding site. The introduction of a premature stop codon into the cadherin coding sequence has also been closely linked to Cry1Ac resistance in *Helicoverpa armigera* (Xu *et al.*, 2005). In this context, it is interesting that three different recessive cadherin mutations were identified in field-collected pink bollworms, *Pectinophera gossypiella* (Morin *et al.*, 2003). Each of the genetic changes resulted in deletion of at least eight amino acids upstream of the putative binding zone. However, resistance only appeared when a combination of any two of the deletion-bearing alleles were present together.

Other mechanisms for resistance to *Bt* toxins besides receptor alterations have been reported, including a deficiency of the initial trypsinergic activation of the protoxin (e.g. Forcada *et al.*, 1996; Oppert *et al.*, 1997; Herrero *et al.*, 2001; Li *et al.*, 2004) and more rapid proteolytic degradation of the activated toxin (Forcada *et al.*, 1996). Recently Li *et al.* (2005) studied the cDNAs coding for three trypsin-like proteinases in a *Bt*-resistant strain of European corn borer which showed reduced activating capacity. In support of this mechanism of resistance it was found that the level of resistance was much lower when the activated protoxin was fed to the larvae. The reduction in trypsinergic activity was attributed to a decrease in gene expression, particularly of one of the three trypsin forms present, OnT23. A detailed study of the midgut proteome of a strain of Indian meal moth resistant to *Bt* revealed a dramatic decrease in chymotrypsin-like activity that led to a decreased rate of activation of the *Bt* protoxin (Candas *et al.*, 2003). However, a number of other structural and functional changes were observed in other parts of the gut systems and it was concluded that a variety of changes in protein expression had resulted from selection and that the resistance to *Bt* was complex and involved multifaceted defences. These included increased glutathione transferases and oxidative capability, and altered energy balance within the midgut epithelial cells that could contribute to increased stress tolerance.

Bs strains are primarily used for mosquito control and their insecticidal toxins have been reviewed by Charles *et al.* (2005). Highly insecticidal strains produce two different toxins, BinA (42 kDa) and BinB (51 kDa), which interact synergistically, as well as other less well-studied cytotoxic proteins. Like the *Bt* endotoxins, the Bin proteins are activated by proteases in the mosquito gut and bind to a receptor on the brush border membrane of gut cells. This receptor has been identified as a glycosylphophatidylinositol-anchored maltase-like enzyme. In contrast to *Bt*, the primary consequence is not to cause general cell lysis, but a more specific action occurs on mitochondria within the target cells which swell and lose respiratory capability. Resistance to *Bs* toxins has been reported several times in field populations of *Culex* species. It is generally monofactorial and recessive. It is not related to the ability to activate the toxin and generally seems to involve changes in receptor responses to the activated toxins. In some cases this results from a loss of binding to the receptor in gut membrane preparations from the resistant strain. In a highly resistant laboratory-selected strain, this apparent loss of receptor binding has been attributed to a mutation in the receptor gene that truncates a hydrophobic tail essential for anchoring the receptor in the cell membrane. Binding still occurs to the receptor secreted into the gut lumen, but it is not then followed by internalization of the toxin. In other cases, the receptor binding characteristics are essentially unaltered in the resistant strains and unidentified changes are presumed to have occurred in the events that follow binding to the receptor (Charles *et al.*, 2005).

3.3. Biotransformation

The metabolic destruction of an insecticide inside the target organism is a common defensive mechanism that leads to a decrease in the duration and intensity of the exposure of the target site and thereby lowers the probability of a lethal outcome. Many insects have developed broad, and often rapidly inducible, defences against potentially toxic xenobiotics that are primarily taken in through the diet, so it is no surprise that these defences may be adapted to act as the second common path to resistance. Three major mechanisms of metabolic transformation of insecticides underlie the vast majority of examples of biotransformation-based resistance: (i) oxidation; (ii) ester hydrolysis; and (iii) glutathione conjugation. Although the products from these reactions are most frequently less toxic than the parent, there are numerous examples of an increase in toxicity as a result of a biotransformation reaction, in which case the insecticide as applied is actually a propesticide. Obviously, an increase in the rate of metabolic conversion in this case should lead to a more toxic result for the insect, and one route to resistance would involve a decrease in the rate of activation. By contrast to resistance arising from site-of-action changes where mutations in the structural genes predominate as an underlying mechanism, biotransformation-based resistance more often involves the overexpression of existing metabolic enzymes by alterations in their regulatory systems and/or through gene duplication. Esterases, cytochrome P450 monooxygenases, and glutathione transferases have all been shown to be significant factors in specific cases of resistance. Epoxide hydrolases, another general class of xenobiotic-metabolizing enzymes, are not known to be involved in insecticide resistance since relatively few insecticides contain epoxide groups and these are sterically hindered, which greatly reduces their susceptibility to hydrolase action (Taniai *et al.*, 2003). Other conjugation reactions such as glucose and sulphate conjugations likewise appear to play little role in known cases of resistance.

Esterases

The majority of widely used insecticides are esters. This includes virtually all carbamates and OPs, most pyrethroids, and other individual compounds such as indoxacarb, methoprene, and similar juvenoids, acequinocyl, spiromesifen and related tetronic acid esters, fluacrypyrim, and bifenazate. In almost all cases, the hydrolysis of the ester group leads to a significant decrease in, or elimination of, toxicity. In just a few cases, ester or amide hydrolysis is an activation reaction; for example, indoxacarb, acequinocyl, or dinitrophenol esters such as dinocap (now largely obsolete) all depend on ester hydrolysis for their toxicity. Consequently, esterase activity often plays a key role in determining the comparative responses and resistance to current insecticides. In insects the primary group of esterases of interest hydrolyse esters of carboxylic acids and they are therefore termed carboxylesterases. The topic of their nature and significance in insecticide toxicology and resistance has been reviewed by Oakeshott *et al.* (2005) and Wheelock *et al.* (2005). Approaches to the classification of carboxylesterases have been described and discussed by Wheelock *et al.* (2005). A useful functional method with special relevance for insecticides was developed by Aldridge (1953) based on the ability of the esterase to either hydrolyse OPs (type A) or to become inhibited by them (type B).

Several structural features of the substrate affect the rates of ester hydrolysis. In addition to electronic effects that affect electron density on the carbonyl carbon atom and influence its susceptibility to nucleophilic attack, esterases are quite sensitive to steric effects in their substrates. In general, esters with larger, bulkier groups close to the ester carbonyl group are less open to attack due to steric hindrance (Buchwald, 2001); within the same series, methyl esters tend to be attacked much faster than isopropyl or tert-butyl esters. Examples include the stabilization of JH analogues such as methoprene by changing the methyl ester, as found in JH itself, to an ispropyl ester (Henrick *et al.*, 1973). Similarly, type II pyethroid esters, with an α-CN group adjacent to the ester, are typically less readily hydrolysed than their type I non-cyano analogues and they tend to be more toxic to insects (Elliott and Janes, 1978).

Many carboxylesterases are also capable of hydrolysing amides and thioesters (Satoh and Hosakawa, 1998). Carboxylamide groups are reasonably common among current insecticides and amide hydrolysis may be an important aspect of their toxicology, such as in the detoxication of benzoylphenylurea inhibitors of chitin synthesis or the OP, dimethoate, or in the hydrolytic activation of the systemic OP, acephate; but the enzymes responsible for such amidase activity have not been much studied. One exception is the purification and partial characterization of a carboxylamidase from the fall armyworm (Yu and Nguyen, 1998). In contrast to the multiple carboxylesterases present, only one carboxylamidase was found. This showed both amidase and esterase activity and had a high sensitivity to several OP inhibitors. A similar carboxyamidase activity was described in the hydrolysis of diflubenzuron by Indian meal moth tissues (Greenberg-Levy *et al.*, 1995). However, the role of such enzymes in resistance remains obscure.

Carboxylesterases (B-esterases)

Insects have 30 or more genes that produce esterases that hydrolyse carboxylic acid esters. They are members of the large and versatile family of enzymes that contain the α/β hydrolase fold with a nucleophile–acid–histidine catalytic triad (Oakeshott *et al.*, 1999). They have been divided into subgroups in several ways (Oakeshott *et al.*, 2005; Wheelock *et al.*, 2005). One subgroup that is inhibited by OPs and preferentially hydrolyses aliphatic substrates is often termed carboxylesterases per se, and these, along with the closely related AChEs, are the forms of primary interest in the context of insecticide resistance. They have also been termed aliesterases because of their preference for aliphatic ester sub-strates, or B-esterases in the nomenclature of Aldridge (1953) because of their susceptibility to inhibition by OPs.

Multiple forms of carboxylesterase are therefore present in insects, often with broad and overlapping substrate specificities, and they may be located in the cytosol, mitochondrial, or microsomal subcellular fractions. Although some cysteine carboxylesterases are known, most carboxylesterases are serine es-terases acting via a catalytic triad with serine as the residue that undergoes se-quential acylation and deacylation. The acylated intermediate may be rapidly broken down to complete the hydrolytic reaction or it may be more stable. In the former case, the hydrolytic turnover of the insecticide will be rapid. At the other end of the continuum, no turnover occurs and the enzyme is permanently acylated. As with AChE, this occurs most generally in the reaction with OP and carbamate esters. This limits catalytic activity to one (partial) turnover, but if enough of the enzyme is present, it may act as a significant sink for the insec-ticide on an equimolar basis. As a source of resistance, it appears that car-boxylesterases may act in both these capacities. Examples are known of the enhancement of carboxylesterase activity on insecticides through gene duplica-tion, and of point mutations that alter the enzyme specificity and kinetics to be more favourable for the hydrolysis of the selecting agent.

A primary example of esterase-based resistance, involving a comprehensive range of biochemical, genetic, and population-level studies primarily conduct-ed at Rothamsted (UK), involves the peach-potato aphid, *M. persicae* (reviewed by Devonshire *et al.*, 1998; Oakeshott *et al.*, 2005). Resistance to carbamates and OPs arises because of the extremely elevated levels of either of two closely related carboxylesterases termed E4 and FE4. Overproduction of the esterase arises from a combination of gene duplication and, in the case of E4, altered transcriptional control through gene methylation. Gene duplication levels can reach as high as 80–100 times the wild-type level and the esterase then accounts for as much as 1% of the aphid's body weight (Field *et al.*, 1999; Bizzaro *et al.*, 2005). Rather surprisingly, a loss of methylation reduces rather than elevates esterase expression of E4 genes and leads to reversion of resistance (Field *et al.*, 1999). Methylation status seems less important in the overexpression of FE4 (Bizzaro *et al.*, 2005). These esterases have a very limited capability to hydrolyse OP and carbamate insecticides, but enough enzyme is present to bind toxico-logically significant amounts of the insecticides, or their activation products in the case of OPs, leading to resistance levels up to 500-fold. This resistance mechanism is particularly effective for OPs since some slow turnover of the phosphorylated enzyme occurs in comparison to carbamates, where little actual hydrolysis occurs and resistance levels are much lower. Similar resistance based on gene duplication leading to enhanced carboxylesterase activity has been re-ported in several other species including other aphids, the mosquito (*Culex* species), and the brown rice planthopper (Hemingway *et al.*, 2004; Oakeshott *et al.*, 2005). Remarkably, in one strain of *C. quinquefasiatus* there is at least a 250-fold increase in gene copy number, a 500-fold increased enzyme level, and about an 800-fold level of OP resistance (Mouchès *et al.*, 1986, 1987).

In contrast to esterases acting as a sink absorbing OPs and carbamates, resistance to OPs in higher dipteran flies has been attributed to specific mutations in a structural gene for carboxylesterase which confer a slow but toxicologically important capability to turn over OPs (Oakeshott *et al.*, 2005). This effect was first reported in houseflies in a pioneering study by Oppenoorth and van Asperen (1960), who used biochemical assays to show in several OP-resistant strains that carboxylesterase (aliesterase) activity was reduced and that this was matched by the simultaneous appearance of an ability to slowly hydrolyse paraoxon and diazoxon, which was the basis of resistance to their parents, parathion and diazinon. This was attributed to a mutation in the aliesterase that conferred OP-hydrolysing capability at the same time as reducing aliesterase activity. Their 'mutant aliesterase' hypothesis has subsequently been confirmed and the molecular mechanism has been shown to be a Gly137Asp substitution in the $\alpha E7$ esterase gene that codes for the E3 esterase protein. This mutation lies in the oxyanion hole of the active site which stabilizes the transition states during both acylation and deacylation through hydrogen bonding. The geometry of carboxyacyl and phosphoryl groups differs, and at this location the considerable change in size and charge of the residues could shift hydrolytic activity from carboxyl esters to phosphate esters if the Asp residue activates a water molecule that is oriented to remove the phosphoryl group attached to the serine of the catalytic triad (Newcomb *et al.*, 1997). Although turnover of the phosphorylated enzyme is still quite slow in the mutant form, it appears to be enough to yield a low but appreciable (around tenfold) resistance to a variety of OPs. This mutation particularly increases the rate of hydrolysis of, and resistance to, diethyl OP esters compared with dimethyl OPs. A less common mutation, Trp251Leu, has been identified at a second locus on the same enzyme. This also leads to decreased carboxylesterase activity and increased OP hydrolysis, although, in this case, dimethyl OP esters are hydrolysed faster than diethyl esters. The same pair of mutations has also been reported in the sheep blowfly, *Lucilia cuprina* (Newcomb *et al.*, 2005), resulting in resistance either to diethyl or dimethyl OP esters depending on the locus of the mutation. In some cases, gene duplication of these mutated esterases has also been observed. This allows mutations at both loci to be present in an individual, which significantly enhances cross-resistance to OPs.

Finally, in an analogy with the aphid/mosquito pattern of resistance, an *E7* esterase gene is overexpressed in resistant horn flies (Guerrero, 2000) without any mutational change in its properties, and this probably protects the insect by diverting inhibitor from the reaction with AChE. The normal physiological role of the *E7* esterase is not known, but it is very interesting to note in *Lucilia* that these mutations, especially at Gly137, initially led to disturbances of bilateral symmetry in the flies, and that *E7*-mediated resistance only became common in the field when an unknown modifier gene was selected which eliminates this effect (McKenzie and O'Farrell, 1993).

For malathion, and a few structurally related OPs such as phenthoate, that have both phosphate and carboxyl ester groups, resistance through enhanced carboxylesterase activity has been reported widely and found in a broad variety of insects (Oakeshott *et al.*, 2005). Both of the carboxylester moieties of malathion are subject to hydrolysis and, in each case, the result is detoxication.

It is interesting to note that these esterase mechanisms tend to be specific to malathion and do not give broad cross-resistance to other OPs. The molecular nature of the malathion esterases is not well established and probably differs between species, nor is it clear whether the increased hydrolytic activity results from overexpression of the enzyme or mutation of the active site, although indirect evidence generally favours the latter. In the few cases where the mechanism has been investigated in more depth, a common theme is that the replacement of larger amino acid residues in the active site by smaller ones has led to better kinetics in the hydrolysis of the rather bulky malathion molecule (Heidari *et al.*, 2005). Along this line of thought, cross-resistance between malathion and the pyrethroids is often seen. This relationship may arise because both malathion and the pyrethroids have bulky acyl groups so that active site mutations that favour access for one group also favour it for the other. Malathion resistance is also conferred by *E*7/E3 esterase enzymes through carboxylesterase action and these are also increased in the housefly and blowfly with the Trp251Leu mutation, perhaps again because of the reduction in size of this substituent (Campbell *et al.*, 1998). The relationship of steric effects to substrate specificity and catalytic activity in OP hydrolysis has been explored in further depth by Heidari *et al.* (2004, 2005) using site-directed mutagenesis in the *L. cuprina* and *D. melanogaster E*7 orthologous genes. Kinetic considerations here are rather complex since with this mutation the enzyme can successfully attack the malathion, and presumably the malaoxon activation product, at the carboxylester groups, and at the same time can slowly cleave the phosphate ester-leaving group of the malaoxon. Furthermore, the ability to free itself from phosphorylation by malaoxon prevents the irreversible inhibition that eventually prevents further carboxylesterase action in the wild-type enzyme (Campbell *et al.*, 1998). In support of this idea, it was found that resistance is much lower to higher alkyl homologues of malathion, which do permanently inhibit the carboxylesterase. The analogous space-making mutation of Trp to Gly is also found in the αE7 esterase of the hymenopteran parasitoid, *Anisopteromalus calandrae* (Zhu *et al.*, 1999a), but a 30-fold up-regulation of the expression of the esterase also occurs (Zhu *et al.*, 1999b), so clearly both mechanisms may operate in insects, and, in this case, in conjunction. An increased level of E3 esterase expression, rather than a change in the catalytic activity, has also been found to be the basis for enhanced malathion hydrolysis in resistant tarnished plant bugs (Zhu *et al.*, 2004).

Finally, Hartley *et al.* (2006) recently reported mutations in esterase E3 of the 'diazinon (Gly137Asp)'- and 'malathion (Trp251Leu)'-type in *Lucilia sericata* that parallel those in its sibling species, *L. cuprina*, demonstrating a case of parallel evolution. The same substitutions also occur in *Musa domestica*, indicating that the options for mutations leading to effective resistance are highly constrained. Significantly, *L. cuprina* specimens collected before the use of OPs in Australasia lacked evidence of the 'diazinon' mutation but several examples of 'malathion' resistance mutations in esterase E3 were found, indicating that these flies were polymorphic at this locus and some were pre-adapted to malathion. This favoured the rapid onset of resistance once malathion was widely deployed in the field.

Most important pyrethroids have carboxylester groups which, if cleaved, results in a complete loss of toxicity. A number of examples of enhanced esterase activity as a component of resistance to pyrethroids have been reported, as reviewed by Oakeshott *et al.* (2005). However, information on the specific nature of the esterases involved, or on the nature of the genetic changes leading to resistance, is quite fragmentary. Since cross-resistance to other esterase-containing insecticides is not typically seen, it is probable that either the esterases themselves, or the changes in structure or expression occurring under pyrethroid selection, are relatively specific for pyrethroid structures. Multiple esterases may be involved in some cases of resistance since many carboxylesterases seem to have at least some ability to hydrolyse pyrethroids. However, it is important to keep in mind that results from carboxylesterase assays in insect tissue using some common model esterase substrates, such as *p*-nitrophenyl acetate, may have little correlation with their ability to hydrolyse pyrethroids (Wheelock *et al.*, 2005). Although the amount of certain esterase isoforms seems to be increased in many cases of pyrethroid resistance, it is also clear from studies using directed mutagenesis of the E3 esterase orthologues from *L. cuprina* and *D. melanogaster* that a variety of point mutations can considerably alter pyrethroid hydrolysis rates and specificities (Heidari *et al.*, 2005), and both mechanisms may well be involved in specific cases.

The recent neurotoxic insecticide, indoxacarb, also has two methyl carboxylester groups. The hydrolysis of one of these occurs rapidly in many insects and results in an activation product (DCJW) that is responsible for the insecticidal action by blocking VDSCs (Wing *et al.*, 1998). The hydrolysis of the other ester group, attached to the oxadiazine moiety, is a detoxication reaction which is generally very slow in insects, perhaps for reasons of steric hindrance, but the rate of this reaction is considerably increased in populations of the oblique-banded leaf roller that are highly resistant to indoxacarb, leading to the rapid degradation of both indoxacarb and, particularly, DCJW (Ahmad and Hollingworth, unpublished data). Little is yet known about the nature of either of these esterases.

A few carboxylesterase reactions lead to metabolites with enhanced toxicity, so an increase in esterase activity could result in faster activation in a resistant strain and thus to negatively correlated cross-resistance. Only a few examples of this have been reported. These include an increased sensitivity to indoxacarb in pyrethroid-resistant *H. armigera* in Australia with elevated esterase activity (Gunning and Devonshire, 2003), and the synthesis of a series of compounds, particularly fluoroacetate esters, specifically designed to take advantage of the elevated esterase activity in resistant potato-peach aphids, and which are more toxic to the resistant forms (Hedley *et al.*, 1998).

Phosphotriesterases

Calcium-dependent phosphotriesterases (A-esterases in the nomenclature of Aldridge, 1953) rapidly hydrolyse many insecticidal OPs, specifically the more labile phosphate esters, generally cleaving the ester at the bond with the most anhydride character (the so-called leaving group) (Vilanova and Sogorb, 1999). They are also active in hydrolysing aromatic carboxylesters and are susceptible to inhibition by mercurial compounds. These enzymes are common in mammals but, as with fish and birds, they seem not to be a major factor in OP metabolism in many insects and are often undetectable in susceptible strains (Dauterman, 1983; Vilanova and Sogorb, 1999). Other phosphotriesterases contain cobalt or zinc at their active site and these also hydrolyse parent phosphorothionates efficiently. In addition to the slow phosphate ester hydrolysis in flies with the mutant carboxylesterases already described, there are reports that the hydrolysis of OPs can occur in other resistant insects, particularly lepidopterans, for example parathion-resistant *H. virescens* (Konno *et al.*, 1990; Kasai *et al.*, 1992), *H. armigera* (Srinivas *et al.*, 2004), and OP-resistant tufted apple bud moths (Devorshak and Roe, 2001). In a strain of houseflies highly resistant to diazinon, Oi *et al.* (1990) detected phosphotriesterase activity against diazoxon in tissues of the resistant insects but not in the susceptible strain. Unfortunately, the nature of these esterases has not always been established in these cases. In some cases, as with the housefly, the phosphotriesterase activity may represent the activity of a mutant carboxylesterase with increased OP turnover. In other cases it seems clear that a metalloenzyme from a different class that hydrolyses arylesters may be involved. Inhibitor profiles, including sensitivity to mercury compounds, indicate the presence of a critical sulphydryl group at the active site. Devorshak and Roe (2001) characterized a phosphotriesterase from the tufted apple bud moth which hydrolysed methyl paraoxon and was associated with resistance to azinphosmethyl, whose activity was stimulated by Ca^{2+}, Co^{2+}, and Mn^{2+}. A similar enzyme activity was also detected at elevated levels in OP-resistant Colorado potato beetles but not in German cockroaches or tobacco budworms. An OP-hydrolysing aromatic esterase has also been described in *Triatoma infestans* (de Malkenson *et al.*, 1984), and Konno *et al.* (1990) partially purified a methyl paraoxonase from resistant tobacco budworms. In these two cases Co^{2+} and Mn^{2+} activated OP hydrolysis but Ca^{2+} did not. Reports of the involvement of such enzymes in resistance are less common than with the carboxylesterases, but this area is not well defined either in significance or in molecular and genetic terms, and it seems to be worthy of further study using the modern methods of molecular genetics and genomics.

Cytochrome P450 monooxygenases

The multifunctional monoxygensases catalysed by cytochrome P450 (MFOs) comprise easily the most versatile system for the metabolism of insecticides in insects. Not only do these enzymes play a determining role in the toxicity of many pesticides, but their variation is also, in many cases, a key to the development of resistance. The topic of insect cytochrome P450-dependent monooxygenases, including their role in resistance, has been reviewed by Bergé *et al.* (1998), Scott (1999), and Feyereisen (1999, 2005).

The monooxygenase system is driven by NADPH through a flavoprotein, NADPH:cytochrome P450 oxidoreductase. Cytochrome b5 may also be involved in electron transfers with some forms of cytochrome P450. Cytochrome P450 uses these reducing equivalents to split and activate molecular oxygen which is then inserted in a large variety of substrates (XH) according to the overall reaction:

$$XH + O_2 + NADPH + H^+ \rightarrow XOH + H_2O + NADP^+.$$

Important reactions from the insecticidal viewpoint include aliphatic carbon oxidations to yield alcohols, aromatic carbon oxidations to yield phenols, and oxidation at unsaturated carbon links to yield epoxides. Oxygen insertion in the double-bonded thionophosphorus linkage leads to the replacement of sulphur by oxygen, a key activation in the toxicity of many OPs. Oxidation at thioether (thiolo) sulphur atoms leads to sulphoxide and sulphone formation which is also an activation reaction for a significant number of OPs and a few other insecticides. The intermediary oxidation products produced by these oxygenases may be unstable and break down further spontaneously; for example, the oxidation of N-alkyl groups to N-alkylols often is followed by the cleavage of the N–C bond effectively resulting in N-dealkylation. The variety of these reactions and the broad substrate specificities of many P450 oxygenases create a defensive system for many insects of remarkable breadth and versatility.

The superfamily of cytochrome P450 (CYP) genes is very large in insects, with approximately 100 so far characterized in each of several insect genomes. Some have known physiological functions in intermediary metabolism, but others are probably present primarily as defences against the many xenobiotic chemicals to which insects may be exposed. Their high level of expression in the midgut, fat body, and Malpighian tubules is strategic in defending against dietary toxicants and those circulating in the haemolymph. Also, some of the P450 enzymes are readily induced in the presence of specific xenobiotics so that the titre of these defensive enzymes is rapidly increased to handle the threat. It has been suggested that the regulatory elements involved in this induction may also be involved in many cases of elevated P450 activity in resistant insects (Liu and Scott, 1997; Scott, 1999).

Evidence for a common role of increased P450 activities in resistance comes from several lines of investigation and illustrates their involvement in examples of resistance to most classes of insecticides (Feyereisen, 2005). Many studies have been conducted with P450 inhibitors, particularly piperonyl butoxide (PBO), which synergize compounds degraded by P450. A reduction in the level of resistance in synergized insects is generally taken to be a useful diagnostic tool indicating a role for P450 in resistance. Such results must be interpreted with care, and preferably with accompanying mechanistic studies, since PBO is not entirely specific for oxygenase reactions; it may also decrease insecticide penetration rates (Sanchez-Arroyo et al., 2001) and inhibit some esterases with synergistic results (Young et al., 2005). The conclusion that MFOs are a factor in resistance because PBO-induced synergism is observed in resistant strains is only valid if similar levels of synergism are absent in susceptible strains. Finally, it should be noted that such inhibitors may not be equally effective on all isoforms of P450. Model P450 substrates have also been widely employed to determine whether the resistant strain has a higher level of P450-catalysed activity in its tissues than the susceptible strain. This can produce useful evidence, but, again, caution is needed since there are many P450 isoforms present and they vary considerably in their substrate specificities, so that model substrates may not well reflect the key changes in oxidase activity underlying resistance. Metabolic studies with the actual insecticide in question are much more reliable indicators of enhanced oxidative metabolism as a factor in resistance. Recently, genomic methodologies have allowed the direct comparison of P450 gene structures and expression between susceptible and resistant strains. However, this approach too has proved to have its limitations in providing specific explanations for resistance.

Because of the complexity of the P450 system and its regulation, and the difficulties in purifying these enzymes from insects (Scott, 1999), it has not been easy to determine unequivocally the exact change(s) in the monooxygenase system that have led to resistance. However, in the cases so far studied, it appears that overexpression of one or more P450 forms has generally occurred, probably through changes in regulatory elements that lead to enhanced transcription. Gene duplication has not so far been detected. However, one instance of point mutations in a structural P450 gene (*Cyp6a2*) leading to enhanced insecticide degradative ability has been described in the case of PBO-suppressible resistance to DDT in a laboratory-selected strain of *D. melanogaster* (RDDT). Overexpression of two forms of P450 (CYP6A2 and CYP4E2) was detected in this strain, but, in addition, a trio of point mutations in *Cyp6a2* enhanced the rate of DDT oxidation approximately tenfold in comparison to the native enzyme. All three mutations were necessary together for maximum activity (Bergé *et al.*, 1998; Amichot *et al.*, 2004).

The current difficulties and progress in unequivocally assigning specific changes in P450 activity to resistance are described in detail by Feyereisen (2005). Unlike the situation with target-site resistance such as the kdr, *ACE*, or *Rdl*, where a very limited number of homologous mutations are found across a broad range of insects, it appears with P450 that a range of forms of the enzyme may be involved among different species, although knowledge outside the *Diptera* is relatively limited. The current challenges in identifying specific P450 changes to explain resistance are illustrated by a long series of studies on the Rutgers strain of housefly which shows a high (roughly 100-fold) level of resistance to the OP, diazinon (Scott, 1999; Feyereisen, 2005). Biochemical studies clearly have shown an enhanced oxidative metabolic capability for diazinon and its activation product, diazoxon, along with a considerably elevated level of one P450 isoform, CYP6A1. Elevated expression of this isoform is also seen in other resistant housefly strains. However, the *Cyp6a1* gene is located on chromosome 5 whereas resistance is unequivocally associated with chromosome 2 (Plapp, 1984). This paradox may be resolvable if the product of the gene on chromosome 2 is a *trans*-acting factor that controls expression of the *Cyp6A1* gene on chromosome 5.

It has been suggested by Feyereisen (2005) that loss of function mutations in *trans*-acting suppressors may commonly underlie the increase in P450 activity associated with resistance. It is likely that these *trans*-acting factors have pleiotropic effects that make analysis of the key changes leading to resistance more difficult. In fact, Plapp (1976) developed a hypothesis that several separate metabolic mechanisms involved in resistance, including P450 monooxygenases, glutathione *S*-transferases (GSTs), and enhanced phosphotriesterase activity, were closely linked and could be controlled by a common element on chromosome 2 in houseflies. No identity for the product of this master switch has so far been advanced. However, Sabourault *et al.* (2001) suggested that the

Gly137Asp mutation in a carboxylesterase (aliesterase) gene (MdαE7) on chromosome 2 was the key event leading to overexpression of CYP genes. This is in addition to its role in converting the carboxylesterase activity to a weak phosphotriesterase action. They postulated (but did not isolate or identify) a repressor molecule produced by the wild-type esterase that was inactive or lacking in the mutant flies. It would be remarkably fortuitous that loss of MdαE7 carboxylesterase activity both leads to an increased ability to hydrolyse the activation product of diazoxon and also, through an independent regulatory activity, turns on production of a P450 that predominantly destroys the parent diazinon.

However, the Gly137Asp mutation is not universally responsible for elevated P450 levels in resistant houseflies. Scott and Zhang (2003) sequenced the critical 700 bp regions of the Md E7 genes in three additional housefly strains (LPR, NG98, and YPER) that exhibit multi-resistance to insecticides through elevated oxidase activity. Both Cyp6a1 and Cyp6d1 were overexpressed in LPR; the specific forms enhanced in the other strains are not described. YPER was heterozygous for the Gly137Asp mutation but it was absent in the other two strains; however these two strains had MdαE7 alleles carrying mutations at another locus (Trp251Leu) that is also known in the sheep blowfly to lead to decreased aliesterase activity and to underlie some types of OP resistance (Campbell et al., 1998). Thus, as long as the presence of these mutations in the heterozygous state is sufficient to affect P450 regulatory activity, the absence of Gly137Asp does not invalidate the hypothesis of an aliesterase/P450 regulatory relationship. In the Rutgers strain, a factor on chromosome 2 also controls expression of Cyp12a1 and Gst-1. These may be pleiotropic effects of the same regulatory agent produced by MdαE7 and could also be involved in the metabolism of diazinon and/or diazoxon. A parallel loss of trans-repression of P450 expression in insecticide-resistant strains probably also operates in D. melanogaster (Feyereisen, 2005).

Scott (1999) and Feyereisen (2005) have reviewed research on a second housefly strain (LPR) that is highly resistant to pyrethroids, with increased oxidative degradation as the major mechanism. In this case it appears that the P450 form involved is CYP6D1, which converts cypermethrin to its 4 -hydroxy metabolite. Resistance is much lower to pyrethroids lacking this hydroxylatable phenoxybenzyl group. CYP6D1 mRNA is overexpressed by about tenfold in the LPR strain due to enhanced transcription, which may involve both cis- and trans-acting mechanisms. The neighbouring Cyp6d3 gene is also strongly overexpressed in this strain (Kasai and Scott, 2001), but its role in resistance is not yet clear. Complicating things further, Shono et al. (2002) examined two additional strains of houseflies from different backgrounds but all selected in the laboratory for resistance to permethrin and all having an enhanced oxidative activity for this compound. They concluded that the genetic basis for this resistance varied between the three strains, suggesting that each selection process had led to a different mechanism of enhanced oxidative metabolism.

By comparison with the housefly studies, the advanced knowledge of the genome in D. melanogaster has allowed greater clarity in defining the role of a monooxygenase at the DDT-R locus that confers resistance to a range of insecticides (Daborn et al., 2001, 2002; Le Goff et al., 2003; reviewed by

Ffrench-Constant *et al.*, 2004). This resistance is characterized by an elevated level of P450 monooxygenase activity. Screening against a microarray containing probes for all 90 *Drosophila* P450 genes showed that only one form of P450, CYP6G1, was overexpressed in the resistant insects compared with susceptibles. Its level of expression was ten to 100 times higher than that in susceptible strains. This assignment is in keeping with the known locus of the resistance gene from the fine-scale mapping of resistance to a region on the polytene chromosome also known to contain *Cyp6g1*. Interestingly, a survey of 20 resistant *Drosophila* strains collected worldwide revealed the same overexpression of *Cyp6g1* in each case. The discovery of the altered expression of the same P450 in all these wild strains stands in clear contrast to the multiple different P450 changes found in resistant strains selected with DDT under laboratory pressure (Feyereisen, 1999). Construction of transgenic insects containing *Cyp6g1* under a tubulin promoter showed that the overexpression of this P450 form conferred moderate resistance in the larvae to a broad range of insecticides including DDT, the chitin-synthesis inhibitor, lufenuron, and several neonicotinoids (Le Goff *et al.*, 2003). It is interesting to note that the survey of the 20 resistant strains found in every case that a transposable element had inserted in the 5′ end of the gene, which may indicate a common origin for these strains that are now found worldwide. A close parallel has been reported in *D. simulans,* where resistance is also associated with overexpression of the orthologous P450 gene and this, too, contains a transposable element insertion (Schlenke and Begun, 2004). One other example of transposon action leading to resistance to *Bt* has already been described (Gahan *et al.*, 2001). These examples raise an intriguing question of the possible role of transposable elements in altering gene expression as a source of resistance, and as a potential factor increasing the frequency of mutations that can lead to resistance in natural populations (Wilson, 1993; Ffrench-Constant *et al.*, 2004, 2006).

Recently, parallel studies were conducted with a field-collected strain of *D. melanogaster* from Australia which was shown to have modest resistance to lufenuron (Bogwitz *et al.*, 2005). In this case, a different P450 gene, *Cyp12a4*, was shown to be specifically overexpressed rather than *Cyp6g1*. Transgenic larvae expressing this gene in the midgut and Malpighian tubules showed enhanced tolerance for lufenuron. This form of P450 is unusual in having a mitochondrial rather than microsomal location in the cell, which raises an issue for the common use of microsomal fractions of cells to assay for P450 activity in resistance. *Cyp6g1* confers a much broader spectrum of resistance to other insecticides than *Cyp12a4*, which may explain why it is the most common form found worldwide in resistant populations. In none of these cases of field-collected strains resistant to lufenuron is there any reason to suppose that they have been previously exposed to this compound, so lufenuron resistance is probably a case of cross-resistance arising from the selection of the monooxygenase by other chemical exposures. However, this case again reveals that resistance to a compound may arise through changes in more than one form of P450 in the same species and that the type of change may vary with the location and past selection history and yield different patterns of cross-resistance. This conclusion is reinforced by the discovery that although *Cyp6g1* is the common form enhanced in resistant *D. melanogaster*, other forms of P450 such as *Cyp6a8* and *Cyp12d1* can be selected for in the laboratory under different DDT selection regimens (Ffrench-Constant *et al.*, 2004).

Many insects beyond these dipterans, including lepidoptera, coleoptera, hemiptera, and dictyoptera, have been shown to develop enhanced oxygenase activity that accompanies, and is probably responsible for, resistance. Despite some knowledge of the specific isoforms that are overexpressed (Feyeresien, 2005), the molecular basis for their increased activity and final proof of their role in resistance to specific compounds are generally lacking. Often, too, the actual metabolites that are produced may not have been identified.

Finally, MFOs are involved in a number of activation reactions in which the oxidized metabolite is the actual toxicant. The oxidation of phosphorothionates to phosphates with much greater AChE-inhibiting capability is the primary example, but others involving such insecticides as diafenthiuron and chlorfenapyr are also known (Hollingworth *et al.*, 1995). In principle, an effective means to develop resistance would be to decrease the rate of activation by loss-of-function mutations or down-regulating expression of the key P450s involved in the activation reaction. A handful of examples of this have been described, for example with methyl parathion in tobacco budworms (Konno *et al.*, 1989), parathion in the Western corn rootworm (Miota *et al.*, 1998), and diazinon in houseflies (Oi *et al.*, 1993), but they are quite uncommon, perhaps because activation can be carried out by several or many P450 isoforms with differing regulation, or because the loss of P450 activity is deleterious to survival. Conversely, an increased P450 activity in a resistant species could lead to enhanced sensitivity to other compounds activated by the same enzyme, causing negatively correlated cross-resistance. This relationship is complex with OPs since MFOs are important both in the activation and inactivation of phosphorothionates and, sometimes, in the inactivation of their phosphate analogues. However, the enhanced sensitivity to triazaphos in *H. armigera* strains from West Africa has been attributed to an elevated monooxygenase activity that also confers resistance to pyrethroids, which illustrates the possibility (Martin *et al.*, 2003). Similar examples involve increased sensitivity to the propesticide, chlorfenapyr, in insects resistant to other insecticides through increased oxidative capability including the tobacco budworm (Pimprale *et al.*, 1997) and, probably, the horn fly (Sheppard and Joyce, 1998).

Glutathione S-transferases

The GSTs are a large group of enzymes that enhance the reaction of the cysteine sulphydryl group of the tripeptide glutathione (GSH) with xenobiotics. The sulphydryl group of GSH is a soft nucleophile that reacts with electrophilic (electron-deficient) sites on xenobiotics leading to the formation of GSH conjugates. These conjugates are more readily excreted than the parent insecticide, and typically are less toxic, although some examples of xenobiotic activation by GSH conjugation are known. The general properties and toxicological relevance of

GSTs have been reviewed by Eaton and Bammler (1999), Sheehan *et al.* (2001), and Hayes *et al.* (2005). Typical electrophiles undergoing GST-catalysed conjugations are the methyl groups of OP esters, halogens or nitro groups in aromatic rings that are activated by the presence of additional electron-withdrawing groups, and, less commonly, α,β-unsaturated ketones, quinones, and epoxides. The dehydrochlorination of DDT to DDE is an unusual example of GST activity in that no intermediate GSH conjugate has been found. Presumably it is too unstable to isolate. Lindane similarly undergoes GSH-dependent conjugations. A critical feature of the mechanism of most GSTs is the activation of the GSH sulphydryl to a higher level of nucleophilicity by the action of a nearby residue with a hydroxyl group (either tyrosine or serine). Many GST-catalysed reactions also occur at a slow rate spontaneously in the absence of the enzyme, but the enzymatic activation of the sulphydryl group accelerates this considerably.

It is becoming clear that insect GSTs may have several different roles in insecticide action and resistance, and these have been reviewed by Yu (1996), Enayati *et al.* (2005), and Ranson and Hemingway (2005). Only the conjugation reactions of an insecticide with GSH are covered in this section; the role of GSTs in antioxidant reactions and in insecticide sequestration is described later.

Although a group of microsomal GSTs exists in insects, the transferases of greatest toxicological interest are soluble. These are relatively small (50–55 kDa) proteins with a dimeric (either hetero or homo) structure. Based on genomic analysis from *Anopheles gambiae* and *D. melanogaster*, six distinct classes of GST, named by Greek letters, have been identified with a total of 30–40 putative GST subunit genes (Ranson and Hemingway, 2005). The nomeclature for GSTs in insects has been revised (Chelvanayagam *et al.*, 2001) and that system is used here. Four of the six classes are common to other organisms including mammals, but two appear to be specific to insects. Multiple enzyme forms occur within each class and have differences in substrate specificity, though often with considerable overlap between forms. The subunit combinations existing in insects are probably within-class specific, but so far are undefined. The two insect-specific classes of GST (δ and ε) predominate in the number of genes present in *A. gambiae* and *D. melanogaster*. Most of the former class readily form GSH conjugates with 1-chloro-2,4-dinitrobenzene (CDNB), but the latter do not. Both classes are centrally involved in reactions with insecticides but there is considerable divergence between the two species in these enzymes and few orthologous structures exist. The ω and σ classes of GST may be involved in protection against reactive oxygen intermediates, while the physiological role of the other classes (θ and ζ) remains unclear. Members of these other classes are also involved in insecticide metabolism, but this role is not fully defined. Two heterodimeric δ class forms of GST (GSTD1–3 and GSTD1–4), formed from alternative splicing products of a gene from *Anopheles dirus* and known to dehydrochlorinate DDT, have been crystallized and their crystal structures determined by Oakley *et al.* (2001). The homodimeric isoform GSTD5–5 has also been crystallized and structurally characterized (Udomsinprasert *et al.*, 2005). Some GSTs are rapidly inducible by enhanced transcription after exposure of the insect to xenobiotics, including both phytochemicals and pesticides, but whether these mechanisms bear any relationship to the genetically based increase in GST activity in resistant forms is unknown (Yu, 1996).

The specific role of the multiple forms is not known and presumably the typical model substrates used to assess the activity of these enzymes in susceptible and resistant species, such as CDNB, 1,2-dichloro-4-nitrobenzene, ethacrynic acid, and monochlorobimane (Nauen and Stumpf, 2002; Yu, 2002), assay a mixture of these isoforms. p-Nitrophenyl acetate is also a substrate for GSTs and it is easy to confuse this activity with that of esterase hydrolysis of the same substrate. Several inhibitors for GST are known (Ranson and Hemingway, 2005), but these have rarely been used as synergists to inhibit GST in insects in vivo and their toxicity, potency, and selectivity for GST isoforms in vivo are undetermined. Attempts to assess the importance of GST resistance in vivo have mainly employed GSH-reactive compounds such as diethyl maleate or N-ethylmaleimide to deplete GSH, which probably has multiple adverse effects on cells. Thus there is not yet a simple, specific, and clearly effective way to demonstrate that GSTs are involved in resistance in vivo.

Although the GSTs are common enzymes with a widespread distribution across insects and with broad substrate specificities, their significance in resistance is more limited than that of oxygenases and esterases, and it has primarily been associated with resistance to organochlorines (DDT, lindane) and conjugations with OPs, acting primarily to O-dealkylate methyl or ethyl phosphate esters, but also to O-dearylate some aromatic OP esters at the leaving group. These compounds are now decreasing in use, but some newer insecticides, such as fipronil (Scharf et al., 2000), appear to have electrophilic centres in their structures (and/or metabolites) that would make them susceptible to GST action, so it may be that resistance could eventually arise to these compounds through enhanced GST conjugation.

Generally GST-based resistance is due to an increased amount of enzyme resulting either from gene duplication or, more often, increased transcription rates resulting from the loss of repressors (Hemingway and Ranson, 2000). One mechanism of overexpression of GSTs occurs in resistant species by the loss of the unidentified trans-acting repressor regulatory mechanism already described for P450 monooxygenase. This leads to amplification of δ class GSTs. But other changed regulatory mechanisms may exist (Ranson and Hemingway, 2005) and changes in mRNA stability remain a possibility, as has been observed in Drosophila. Point mutations in the structural genes have not been related to resistance as yet, but it is clear that single amino acid substitutions in GSTs from A. dirus, even at a distance from the active site, can change their conformation, substrate specificity, and their susceptibility to inhibition by pyrethroid insecticides (Wongsantichon et al., 2003; Wongtrakul et al., 2003).

In only a few cases has it yet been possible to associate resistance with an increased activity of specific isoforms of GST. For example, the δ form GSTD5–5 from A. dirus has very high DDT dehydrochlorinase activity but, interestingly, has little activity against other typical GST substrates (Udomsinprasert et al., 2005); however, it is not necessarily the major source of DDT resistance in this strain. In Aedes gambiae, DDT resistance is associated with an ε class subunit gene Gste2 and the homodimer of its product, GSTE2–2, rapidly metabolizes

DDT (Lumjuan *et al.*, 2005). Wei *et al.* (2001) cloned a gene for a θ class GST, *Gstt-6a*, that is overproduced in houseflies highly resistant to OP insecticides. Numerous GSTs are also overproduced in this strain but GSTT-6A carried out GSH conjugations with both methyl parathion and lindane at a high rate compared with other overexpressed forms, and the authors concluded it was probably a key factor in the resistance shown by this strain.

Thus the role of the GST family in insecticide resistance is complex and still quite poorly defined. Multiple isoforms from several classes are involved and their substrate specificities seem variable and hard to predict.

3.4. Other Resistance Mechanisms

In addition to changes in target sites and biotransformation systems, several other biochemical mechanisms may contribute to resistance at a more modest level. Although individually they may be only moderate in their impact on toxicity, they can act as important intensifiers of resistance when combined with the major mechanisms in the same insect. They also tend to be broad-spectrum in their action. Relatively little attention has been paid to these mechanisms and they are still rather poorly understood in many cases.

Decreased rates of penetration

Studies with a variety of insects and insecticides have revealed situations where the uptake of the insecticide applied externally to a resistant strain is slower than that in a susceptible one, as evidenced by a slower rate of appearance of the compound in internal extracts and greater retention on the insect surface. In houseflies the reduced uptake is controlled by a single recessive gene, but the phenomenon is still not fully understood at the biochemical level and it has received relatively little investigation since it was reviewed by Plapp (1976).

This effect of reduced penetration on toxicity is generally quite small, typically in the range of two- or threefold. Although this is not a large effect, it clearly contributes to the overall level of resistance as a subsidiary factor that augments the effects of other mechanisms and is active on a diverse range of insecticides, including reports involving organotins, carbamates, OPs, and avermectins. A slower rate of penetration should lead to a lowered internal concentration of the active toxicant as long as there is concomitant removal of the compound by metabolism, excretion, or permanent sequestration. In genetic studies with the housefly, penetration effects interacted multiplicatively with other resistance genes (Plapp and Hoyer, 1968; Raymond *et al.*, 1989). Reduced uptake is controlled by a gene *Pen* (Sawicki and Lord, 1970), also previously termed *tin* by Hoyer and Plapp (1968) because of its effect on the uptake of organotin pesticides, located on chromosome 3 in the housefly. It has been concluded by Gardiner and Plapp (1997) that an energy-dependent transporter protein which normally moves hydrocarbons from the haemolymph to the outer cuticle may also be involved in transporting lipophilic insecticides outwards, and that changes in the amount or nature of this protein that increase the efficiency of outward movement would result in a decrease in insecticide accumulation internally. This transporter is yet to be identified. Szeicz et al. (1973) have also reported decreased penetration as a factor in the resistance of tobacco budworm larvae to a variety of insecticides. In parallel to the situation in houseflies, Lanning et al. (1996a) proposed a key role for a cuticular P-glycoprotein (P-gp) transporter in the movement of insecticides across the integument and as a factor in their decreased uptake and internal accumulation in resistant tobacco budworms. Treatment of the budworm larvae with quinidine, a P-gp transport inhibitor, was shown to increase the rate of internal accumulation of topically applied thiodicarb in the resistant strain as well as to increase its toxicity. The final link in these relationships is still missing, which is to show that the *Pen* gene, or its homologue in other species, codes for a transporter protein such as a cuticular P-gp.

Removal of the toxicant by efflux pumps

If carriers can move insecticides back to the cuticle from internal sites and thus reduce the overall rate of penetration, could such transporters also act internally to protect critical physiological systems? The action of transporters that remove a broad range of toxic compounds from cells by energy-dependent P-gp transmembrane pumps is well established as a mechanism of antibiotic resistance in bacteria (Langton et al., 2005) and of fungicide resistance in fungi (Nakaune et al., 1998; Andrade et al., 2000). Increased efflux has also been identified as a probable cause of resistance to ivermectin and its relatives in parasitic helminths (Xu et al., 1998; Kerboeuf et al., 2003).

These transporters, as members of the ATP binding cassette (ABC) superfamily, are also present in insects; for example, three P-gp multi-drug resistance genes (mdr) have so far been identified in *D. melanogaster* (Gerrard et al., 1993). Evidence that they can be a factor leading to resistance in insects is fragmentary though suggestive. P-gp pumps have been identified in both the Malpighian tubules and the blood–brain barrier of tobacco hornworm larvae. These remove nicotine taken in through the diet and help to endow this species with nicotine resistance (Murray et al., 1994). In a study with tobacco budworm larvae, Lanning et al. (1996b) showed a several-fold increase in the expression of P-gp, not only in the cuticle but also in the fat body, in several resistant strains. Quinidine, an inhibitor of this pump, synergized the topical toxicity of thiodicarb with a much stronger effect on the resistant strain (12.5-fold compared with 1.8-fold in susceptible insects). Srinivas et al. (2004) have also reported that P-gp expression could be readily detected in multi-resistant larvae of *H. armigera*, but not in susceptible insects, although a variety of other metabolic and target site resistance mechanisms were also present. Buss et al. (2002) reported that in larvae of the *C. pipiens* complex, treatment with another P-gp inhibitor, verapamil, increased the toxicity of cypermethrin, endosulfan, and ivermectin, although in this case the degree of synergism, about twofold, was the same in both susceptible and resistant strains. An important consideration in studies using putative P-gp inhibitors *in vivo* is that it is necessary to be sure they are acting specifically on the drug transporter in causing synergism, since these compounds can also affect other physiological systems in insects. For example, quinidine also inhibits some forms of cytochrome P450 and blocks neuronal sodium and potassium channels among several potentially significant pharmacological actions. This demonstration of a P-gp-specific synergistic action has been approached only in the case of the tobacco budworm study of Lanning *et al*. (1996b).

Binding of the insecticide at toxicologically harmless sites

The binding of toxicants to macromolecules that does not itself result in an adverse response may occur either covalently (generally irreversibly) or non-covalently (generally reversibly). Covalent binding occurs with compounds that are chemically reactive in biological systems such as the carbamates and OPs, or the carbodiimide metabolite produced by the activation of diafenthiuron. The reaction of OPs and carbamates with carboxylesterases, which thereby act as a sink for the active cholinesterase inhibitors, has already been described above as an important source of resistance in homopterans and mosquitoes. It is likely, though undetermined, that other acylatable macromolecules can also react with and remove OPs and carbamates.

Non-covalent binding may also be important in governing the toxicity of pesticides; in vertebrates, for example, serum albumin is well established as an important factor governing the availability of drugs and pesticides. In insects, most of the pyrethroid present internally is adsorbed to 'solids' (Burt *et al*., 1971; Welling, 1977) and is not freely available in solution, and this is probably true for most lipophilic pesticides. Extensive binding to proteins in the haemolymph is commonly observed (Boyer, 1975; Skalsky and Guthrie, 1977). However, there has been little recent study of this topic, and the nature of the binding proteins is not well established although a lipophorin and vitellogenin have been identified in one case (Breton *et al*., 1992). This reversible binding creates a reservoir of the insecticide that modulates its concentration *in vivo* and, assuming reasonable reversibility of binding, could also act as an internal transport system via the haemolyph and move compounds from areas of higher to lower concentration. The contribution, if any, of changes in such binding to resistance is unknown.

Some GSTs are expressed in large amounts, particularly in resistant insects, and in addition to their catalytic functions they also bind high-molecular weight compounds (>400 Da). This activity was initially attributed to a 'ligandin' protein before it was established that the binding was due to GST. The primary ligand-binding site on GST has been identified as a portion of the xenobiotic substrate-binding site (Oakley *et al*., 1999) and many compounds that bind there also inhibit GST activity, including organotins (Henry and Byington, 1976), OPs (Ketterman *et al*., 2001), and pyrethroid insecticides (Ketterman *et al*., 2001; Kostaropoulos *et al*., 2001; Udomsinprasert and Ketterman, 2002; Wongtrakul *et al*., 2003). In this context, it has been suggested that the elevated levels of

GST in resistant insects may contribute to resistance through the increased pyrethroid binding that leads to sequestration of the compound (Grant and Matsumura, 1989; Kostaropoulos et al., 2001), although this was found not to be a resistance mechanism in brown planthoppers resistant to pyethroids through increased GST activity (Vontas et al., 2001).

Downstream effects following the initial attack at the target site

A variable, but possibly substantial, number of events and processes are involved between the initial action of an insecticide at its target site and the final cellular and physiological events that lead irreversibly to death. Since it is generally unclear why insects die after exposure to insecticides, some of these events remain unknown or, at best, are poorly investigated. However, it is not unreasonable to suppose that resistance could arise by an elevated capability to either tolerate or repair the initial damage initiated by the insecticide at the target site. Only a few examples of this are known, but, then, the topic has not been much studied.

One recent example does illustrate that such effects can be a factor in resistance. There are several reports of elevated GST activity in pyrethroid-resistant strains and that pyrethroid insecticides can be synergized by compounds that interfere with the action of GSTs; for example by pretreatment with GST inhibitors such as cibacron blue (Pospischil et al., 1996) or with diethyl maleate which depletes tissue GSH (Ahmad and Hollingworth, 2004). However, there is no indication that the GSTs are involved in the metabolism of most pyrethroids. Possible sequestration effects by ligand binding to GST have been considered, as already described, but another reasonable and novel explanation for this action has been presented by Vontas et al. (2001, 2002) and reviewed by Ranson and Hemingway (2005). They provided evidence that in the brown planthopper, pyrethroids cause important tissue injury by generating reactive oxygen intermediates that can attack vital macromolecules in the cell and thereby cause cell death. This action, involving reactive oxygen species (ROS) such as hydrogen peroxide, superoxide anion, and hydroxyl radicals, is well studied in vertebrates and has been termed 'oxidative stress' (see e.g. Finkel and Holbrook, 2000). Similar processes occur in insects (Le Bourg, 2001). There are many targets in the cell for ROS, but lipids are a particularly sensitive site for peroxidation leading to free radical chain reactions, the formation of highly reactive lipid breakdown products, and, eventually, to membrane destruction. The primary source of the ROS is the mitochondrial respiratory chain, but other oxygen-activating systems, including cytochrome P450, can produce them. Pyrethroids are among the many compounds known to cause oxidative stress in vertebrates (Abdollahi et al., 2004; Prasanthi et al., 2005), although the mechanism is unclear. GSTs and GSH itself play several key roles in combating oxidative stress, particularly through their glutathione peroxidase activity that destroys reactive hydroperoxides, and by conjugation of GSH with the reactive intermediate products of the oxidation of lipids (see e.g. Hayes et al., 2005). These GSH-dependent antioxidative defences are present and important in insects (Parkes *et al.*, 1993) and are inducible in mosquitoes after exposure to peroxides (Ding *et al.*, 2005).

In their studies with the brown planthopper, Vontas *et al.* (2001) found that the resistant insects had elevated levels of GSTs that predominantly showed peroxidase activity, and that these were more effective in decreasing the pyrethroid-initiated lipid peroxidation in the resistant strain compared with the susceptible strain. Pretreatment of the adults with the GST inhibitor, ethacrynic acid, removed most of the GST peroxidase activity, synergized the toxicity of the pyrethroids and largely eliminated resistance. Subsequently, Vontas *et al.* (2002) examined the mechanism of enhancement of GST activity and tentatively attributed it to the overexpression in the resistant strain of one class I form with peroxidase activity (GSTI-1) due to gene duplication. However it is likely that this is not the only overexpressed form of GST in the resistant insects. In this case, the level of pyrethroid resistance was relatively low (about ten- to 20-fold), and it is not clear whether this mechanism can endow high-level resistance on insects, but a strong case has been made for enhanced oxidative defences as a factor in resistance.

Many other insecticides are known to cause significant oxidative stress in higher organisms including organochlorines, OPs, carbamates, and nicotine (Abdollahi *et al.*, 2004; Qiao *et al.*, 2005). A variety of insecticides and acaricides that act as inhibitors of mitochondrial respiration also have the capability to generate ROS and cause oxidative stress (Hollingworth, 2001). Thus it is possible that the toxicity of these compounds may also be dependent to some degree on the GSH/GST status of the insect for protection against ROS and that this may represent a mechanism of resistance with significance well beyond the pyrethroids.

In another possible example of a downstream resistance mechanism, Ferré and Van Rie (2002) speculated that an enhanced ability to repair or replace *Bt* toxin-damaged cells may be a resistance mechanism in *H. virescens*, since the initial midgut damage caused by *Bt* consumption appeared to be comparable in both resistant and susceptible strains. Also, using proteomic analysis, a complex of midgut adaptations that is probably related to stress responses was discovered in *Bt*-resistant Indian meal moths (Candas *et al.*, 2003). Finally, Griffitts and Aroian (2005) discuss a number of downstream inducible protective responses against *Bt* intoxication that could be enhanced genetically to provide inheritable resistance, although these have not yet been the subject of much study.

3.5. Understanding the Integrated Effect of Multiple Mechanisms: the Pharmacokinetics and Pharmacodynamics of Resistance

In the past two decades or so, new methods of studying biochemical mechanisms of resistance and their underlying molecular genetics have advanced our understanding of this aspect of the basis of resistance in a remarkable way. They have produced powerful tools for studying the population genetics of resistance as a microevolutionary phenomenon and for practical use in resistance management and the search for new insecticides. By contrast, the understanding of resistance at the whole insect level has not advanced comparably.

While the discovery of a less-sensitive target site or strongly enhanced enzymatic degradation can provide a plausible explanation for a given example of resistance, the relationship between changes in biochemical parameters and increased resistance is not always so easy to assess, particularly quantitatively and where multiple resistance mechanisms exist. Past approaches to assessing the interactions of multiple resistance mechanisms have included the breeding of a variety of polygenic resistant strains from ones with known single resistance factors. For example, using this approach in the case of the housefly, reduced penetration was shown to interact multiplicatively with other resistance genes (Plapp and Hoyer, 1968; Sawicki, 1973; Raymond et al., 1989), but other interactions may be more complex (Plapp, 1976).

In certain situations, genetic technology now allows an unequivocal assessment of the role of a specific mechanism in resistance, such as high resistance due to the complete loss of a sensitive target site of action by a null mutation, as described in the Introduction. The development of transgenic insects in which individual metabolic enzymes are inserted under promoters that allow the control of their expression can also provide direct evidence of their significance in resistance to different insecticides. For example, Daborn et al. (2002) and Le Goff et al. (2003) transferred the Cyp6g1 gene to D. melanogaster and showed that the specific overexpression of this gene led to resistance to DDT and a range of other insecticides including neonicotinoids and insect growth regulators. They also point out that the use of microarray analysis to determine the expression levels of insecticide metabolic enzymes can greatly enhance the analysis of the role of different types of biotransformation systems and of specific isoforms and combinations of isoforms in individual cases of resistance. Despite such advances, currently limited to a small range of experimental insects, a full analysis of resistance remains challenging in the common case where multiple independent resistance mechanisms are present in the same insect, or in other situations where a mutation may lead to the pleiotropic change in expression of several metabolic enzymes or to a mutant receptor subunit that may be present in several different receptor combinations.

As a final step in analysing the interactions of resistance mechanisms in the whole organism, comparative physiologically based pharmacokinetic/pharmacodynamic (PBPK/PD) modelling should be considered. This technique, quite widely used in pharmacological and toxicological research with vertebrates, is poorly developed in insects despite several pioneering attempts with different levels of sophistication (see e.g. Hollingworth, 1971; Welling, 1977, 1979; Ford et al., 1981; Lagadic et al., 1993; Chalvet-Monfray et al., 1998; Greenwood et al., 2002, 2007). This approach should be particularly useful in assessing the relative significance of each component where multiple resistance mechanisms are present and also in identifying key gaps that clearly still exist in our understanding of the complex, and often non-linear and interactive, processes of insecticide action. For example, the uptake/penetration of topically applied insecticides is typically a multiphase process. The rate is not linearly related to dose over a wide range of doses, and it may be affected by the rate of internal removal of the insecticide, for example by metabolism or tissue binding, which steepens the concentration gradient controlling inward diffusion. Also, enzymes are saturable, and, over time, many are inducible. Co-substrates such as glutathione may be depleted at high doses, product inhibition of key enzymes may occur (not uncommon with cytochrome P450 reactions), receptor levels can be up- and down-regulated, and a variety of other adaptations to stress can be occur during the poisoning process. Thus the results of PBPK/PD studies are not always easily predictable or intuitively obvious.

Only a handful of PBPK modelling studies have addressed resistance in insects, and, by necessity, these generally were pursued at a low level of complexity. Welling *et al.* (1983) modelled malaoxon toxicity in susceptible and resistant houseflies, and developed a model that fitted the experimental data with the major factor in resistance being an enhanced rate of oxidative destruction of the insecticide. Oi and colleagues (Oi *et al.*, 1993; Oi, 1996) analysed the pharmacokinetics of diazinon and diazoxon in susceptible and resistant houseflies. Multiple metabolic effects were identified as interacting to reduce the internal levels of the activation product, diazoxon. The enhanced degradation of diazoxon was a primary factor, with an increased rate of destruction of diazinon and slower activation to diazoxon acting as subsidiary mechanisms in the resistant strain. Hollingworth (1971) modelled the penetration and metabolism of fenitrothion in susceptible and resistant houseflies and concluded that the primary mechanism of resistance was enhanced hydrolysis of the oxidative activation product, fenitroxon, with a smaller contribution from slower penetration in the resistant strain. Finally, Lagadic *et al.* (1993) studied the pharmacokinetics of cyfluthrin in susceptible and resistant *Spodoptera littoralis* and concluded that although differences existed between the strains in the distribution of the compound and the loss through regurgitation in the susceptible insects, the primary reason for resistance lay in an insensitive (kdr-like) site of action.

For reasonable accuracy, such studies require the collection of a considerable amount of data and an understanding of each of the critical steps in the chain of events from first contact with the insecticide to the final physiological disruptions leading to irreversible injury and death. The fact that these data and underlying knowledge are often too deficient to develop a realistic and complete pharmacokinetic/dynamic model suggests that there is much yet to be learnt about insecticide action and resistance at the whole organism level.

References

Abdollahi, M., Ranjbar, A., Shadnia, S., Nikfar, S. and Rozale, A. (2004) Pesticides and oxidative stress: a review. *Medical Science Monitor* 10, RA141–RA147.

Ahmad, M. and Hollingworth, R.M. (2004) Synergism of insecticides provides evidence of metabolic mechanisms of resistance in the obliquebanded leafroller *Choristoneura rosaceana* (Lepidoptera: Tortricidae). *Pest Management Science* 60, 465–473.

Aldridge, W.N. (1953) Serum esterases. 1. Two types of esterase (A and B) hydrolysing p-nitrophenyl acetate, propionate and butyrate, and a method for their determination. *Biochemical Journal* 53, 110–117.

Amichot, M., Tarès, S., Brun-Barale, A., Arthaud, L., Bride, J.-M. and Bergé, J.-B. (2004) Point mutations associated with insecticide resistance in the *Drosophila* cytochrome P450 *Cyp6a2* enable DDT metabolism. *European Journal of Biochemistry* 271, 1250–1257.

Andrade, A.C., Del Sorbo, G., Van Nistelrooy, J.G.M. and De Waard, M.A. (2000) The ABC transporter AtrB from *Aspergillus nidulans* mediates resistance to all major classes of fungicides and some natural toxic compounds. *Microbiology* 146, 1987–1997.

Anthony, N., Unruh, T., Ganser, D. and Ffrench-Constant, R.H. (1998) Duplication of the Rdl GABA receptor in an insecticide-resistant aphid, *Myzus persicae*. *Molecular and General Genetics* 260, 165–177.

Baxter, S.W., Zhao, J.Z., Gahan, L.J., Shelton, A.M., Tabashnik, B.E. and Heckel, D.G. (2005) Novel genetic basis of field-evolved resistance to Bt toxins in *Plutella xylostella*. *Insect Molecular Biology* 14, 327–334.

Bergé, J.B., Feyereisen, R. and Amichot, M. (1998) Cytochrome P450 monooxygenases and insecticide resistance in insects. *Philosophical Transactions of the Royal Society of London. Series B, Biological Sciences* 353, 1701–1705.

Bizzaro, D., Mazzoni, E., Barbolini, E., Giannini, S., Cassanelli, S., Pavesi, F., Cravedi, P. and Minacardi, G.C. (2005) Relationship among expression, amplification, and methylation of FE4 esterase genes in Italian populations of *Myzus persicae* (Sulzer) (Homoptera: Aphididae). *Pesticide Biochemistry and Physiology* 81, 51–58.

Bloomquist, J.R. (2001) GABA and glutamate receptors as biochemical sites for insecticide action. In: Ishaaya, I. (ed.) *Biochemical Sites of Insecticide Action and Resistance*. Springer, Berlin, pp. 17–41.

Bloomquist, J.R. (2003) Chloride channels as tools for developing selective insecticides. *Archives of Insect Biochemistry and Physiology* 54, 145–156.

Bogwitz, M.R., Chung, H., Magoc, L., Rigby, S., Wong, W., O'Keefe, M., McKenzie, J.A., Batterham, P. and Daborn, P.J. (2005) *Cyp12a4* confers lufenuron resistance in a natural population of *Drosophila melanogaster*. *Proceedings of the National Academy of Sciences USA* 102, 12807–12812.

Bourguet, D., Guillemaud, T., Chevillon, C. and Raymond, M. (2004) Fitness costs of insecticide resistance in natural breeding sites of the mosquito *Culex pipiens*. *Evolution* 58, 128–135.

Boyer, A.C. (1975) Sorption of tetrachlorvinphos insecticide (Gardona) to the hemolymph of *Periplaneta americana*. *Pesticide Biochemistry and Physiology* 5, 135–141.

Bravo, A., Soberón, M. and Gill, S.S. (2005) *Bacillus thuringiensis*: mechanisms and use. In: Gilbert, L.I., Iatrou, K. and Gill, S.S. (eds) *Comprehensive Insect Molecular Science*, Vol. 6. Elsevier, Oxford, UK, pp. 175–205.

Breton, P., Vanderhorst, D.J., Vandoorn, J.M. and Beenakkers, A.M.T. (1992) Binding of lindane to locust hemolymph proteins. *Comparative Biochemistry and Physiology. Part C, Pharmacology, Toxicology & Endocrinology* 101, 137–142.

Buchwald, P. (2001) Structure–metabolism relationships: steric effects and the enzymatic hydrolysis of carboxylic esters. *Mini Reviews in Medicinal Chemistry* 1, 101–111.

Buckingham, S.D. and Sattelle, D.B. (2005) GABA receptors of insects. In: Gilbert, L.I., Iatrou, K. and Gill, S.S. (eds) *Comprehensive Insect Molecular Science*, Vol. 5. Elsevier, Oxford, UK, pp. 107–142.

Buckingham, S.D., Biggin, P.C., Sattelle, B.M., Brown, L.A. and Sattelle, D.B (2005) Insect GABA receptors: splicing, editing, and targeting by antiparasitics and insecticides. *Molecular Pharmacology* 68, 942–951.

Burt, P.E., Lord, K.A., Forrest, J.M. and Goodchild, R.E. (1971) The spread of topically applied pyrethrin I from the cuticle to the central nervous system of the cockroach, *Periplaneta americana*. *Entomologia Experimentalis et Applicata* 14, 255–269.

Buss, D.S., McCaffery, A.R. and Callaghan, A. (2002) Evidence for P-glycoprotein modification of insecticide toxicity in mosquitoes of the *Culex pipiens* complex. *Medical and Veterinary Entomology* 16, 218–222.

Campbell, P.M., Newcomb, R.D., Russell, R.J. and Oakeshott, J.G. (1998) Two different amino acid substitutions in the ali-esterase, E3, confer alternative types of organophosphorus insecticide resistance in the sheep blowfly, *Lucilia cuprina*. *Insect Biochemistry and Molecular Biology* 28, 139–150.

Candas, M., Loseva, O., Oppert, B., Kosaraju, P. and Bulla, L.A. Jr (2003) Insect resistance to *Bacillus thuringiensis* – alterations in the Indianmeal moth larval gut proteome. *Molecular and Cellular Proteomics* 2, 19–28.

Chalvet-Monfray, K., Belzunces, L.P. and Auger, P. (1998) A theoretical study of discriminating parameters in metabolic resistance to insecticides. *Pesticide Science* 52, 354–360.

Charles, J.-F., Darboux, I., Pauron, D. and Nielsen-Leroux, C. (2005) Mosquitocidal *Bacillus sphaericus*: toxins, genetics, mode of action, use and resistance mechanisms. In: Gilbert, L.I., Iatrou, K. and Gill, S.S. (eds) *Comprehensive Insect Molecular Science*, Vol. 6. Elsevier, Oxford, UK, pp. 207–232.

Charpentier, A. and Fournier, D. (2001) Levels of total acetylcholinesterase in *Drosophila melanogaster* in relation to insecticide resistance. *Pesticide Biochemistry and Physiology* 70, 100–107.

Chelvanayagam, G., Parker, M.W. and Board, P.G. (2001) Fly fishing for GSTs: a unified nomenclature for mammalian and insect glutathione transferases. *Chemico-biological Interactions* 133, 256–260.

Clark, J.M. and Yamaguchi, I. (2002) *Agrochemical Resistance: Extent, Mechanism, and Detection*. ACS Symposium Series No. 808. American Chemical Society, Washington, DC.

Clark, J.M., Scott J.G., Campos, F. and Bloomquist, J.R. (1995) Resistance to avermectins: extent, mechanisms, and management implications. *Annual Review of Entomology* 40, 1–30.

Cornel, A.J., Stanich, M.A., McAbee, R.D. and Mulligan, F.S. (2002) High level methoprene resistance in the mosquito *Ochlerotatus nigromaculis* (Ludlow) in Central California. *Pest Management Science* 58, 791–798.

Daborn, P., Boundy, S., Yen, J., Pittendrigh, B. and Ffrench-Constant, R. (2001) DDT resistance in *Drosophila* correlates with Cyp6g1 over-expression and confers cross-resistance to the neonicotinoid imidacloprid. *Molecular Genetics and Genomics* 266, 556–563.

Daborn, P.J., Yen, J.L., Bogwitz, M.R., Le Goff, G., Feil, E., Jeffers, S., Tijet, N., Perry, T., Heckel, D., Batterham, P., Feyereisen, R., Wilson, T.G. and Ffrench-Constant, R.H. (2002) A single P450 allele associated with insecticide resistance in *Drosophila*. *Science* 297, 2253–2256.

Dauterman, W.C. (1983) Role of hydrolases and glutathione-S-transferases in insecticide resistance. In: Georghiou, G.P. and Saito, T. (eds) *Pest Resistance to Pesticides: Challenges and Prospects*. Plenum, New York, pp. 229–248.

de Malkenson, N.C., Wood, E.J. and Zerba, E.N. (1984) Isolation and characterization of an esterase of *Triatoma infestans* with a critical role in the degradation of organophosphorus esters. *Insect Biochemistry* 14, 481–486.

Denholm, I., Pickett, J.A. and Devonshire, A.L. (1999) *Insecticide Resistance: From Mechanisms to Management*. CAB International, Wallingford, UK.

Dent, J.A., Smith, M.M., Vassilatis, D.K and Avery, L. (2000) The genetics of ivermectin resistance in *Caenorhabditis elegans*. *Proceedings of the National Academy of Sciences USA* 97, 2674–2679.

Devonshire, A.L., Field, L.M., Foster, S.P., Moores, G.D., Williamson, M.S. and Blackman, R.L. (1998) The evolution of insecticide resistance in the peach-potato aphid, *Myzus persicae*. *Philosophical Transactions of the Royal Society of London. Series B, Biological Sciences* 353, 1677–1684.

Devorshak, C. and Roe, R.M. (2001) Purification and characterization of a phosphoric triester hydrolase from the tufted apple bud moth, *Platynota idaeusalis* (Walker). *Journal of Biochemical and Molecular Toxicology* 15, 55–65.

Dhadialla, T.S., Carlson, G.R. and Le, D.P. (1998) New insecticides with ecdysteroidal and juvenile hormone activity. *Annual Review of Entomology* 43, 545–569.

Dhadialla, T.S., Retnakaran, A. and Smagghe, D. (2005) Insect growth- and development-disrupting insecticides. In: Gilbert, L.I., Iatrou, K. and Gill, S.S. (eds) *Comprehensive Insect Molecular Science*, Vol. 6. Elsevier, Oxford, UK, pp. 55–115.

Ding, Y., Hawkes, N., Meredith, J., Eggleston, P., Hemingway, J. and Ranson, H. (2005) Characterization of the promoters of epsilon glutathione transferases in the mosquito *Anopheles gambiae* and their response to oxidative stress. *Biochemical Journal* 387, 879–888.

Dong, K. (1993) Molecular characterization of knockdown (kdr)-type resistance to pyrethroid insecticides in the German cockroach (*Blattella germanica* L.). PhD dissertation, Cornell University, Ithaca, New York.

Dong, K. (2007) Insect sodium channels and insecticide resistance. *Invertebrate Neuroscience* 7, 17–30.

Dong, K. and Scott, J.G. (1994) Linkage of the kdr-type resistance locus to the sodium channel gene in German cockroaches. *Insect Biochemistry and Molecular Biology* 24, 647–654.

Eaton, D.L. and Bammler, T.K. (1999) Concise review of the glutathione S-transferases and their significance to toxicology. *Toxicological Sciences* 49, 156–164.

Elliott, M. and Janes, N.F. (1978) Synthetic pyrethroids – a new class of insecticide. *Chemical Society Reviews* 7, 473–505.

Enayati, A.A., Ranson, H. and Hemingway, J. (2005) Insect glutathione transferases and insecticide resistance. *Insect Molecular Biology* 14, 3–8.

Feng, G., Deak, P., Chopra, M. and Hall, L.M. (1995) Cloning and functional analysis of TipE, a novel membrane protein that enhances *Drosophila* para sodium channel function. *Cell* 82, 1001–1011.

Ferré, J. and Van Rie, J. (2002) Biochemistry and genetics of insect resistance to *Bacillus thuringiensis*. *Annual Review of Entomology* 47, 501–533.

Feyereisen R. (1995) Molecular biology of insecticide resistance. *Toxicology Letters* 82/83, 83–90.

Feyereisen, R. (1999) Insect P450 enzymes. *Annual Review of Entomology* 44, 507–533.

Feyereisen, R. (2005) Insect cytochrome P450. In: Gilbert, L.I., Iatrou, K. and Gill, S.S. (eds) *Comprehensive Insect Molecular Science*, Vol. 4. Elsevier, Oxford, UK, pp. 1–77.

Ffrench-Constant, R.H., Anthony, N., Aronstein, K., Rocheleau, T. and Stilwell, G. (2000) Cyclodiene insecticide resistance: from molecular to population genetics. *Annual Review of Entomology* 45, 449–466.

Ffrench-Constant, R.H., Daborn, P.J. and Le Goff, G. (2004) The genetics and genomics of insecticide resistance. *Trends in Genetics* 20, 163–170.

Ffrench-Constant, R., Daborn, P. and Feyereisen, R. (2006) Resistance and the jumping gene. *BioEssays* 28, 6–8.

Field, L.M., Blackman, R.L., Tyler-Smith, C. and Devonshire, A.L. (1999) Relationship between amount of esterase and gene copy number in insecticide-resistant *Myzus persicae* (Sulzer). *Biochemical Journal* 339, 737–742.

Finkel, T. and Holbrook, N.J. (2000) Oxidants, oxidative stress and the biology of ageing. *Nature* 408, 239–247.

Flatt, T., Tu, M.P. and Tatar, M. (2005) Hormonal pleiotropy and the juvenile hormone regulation of *Drosophila* development and life history. *BioEssays* 27, 999–1010.

Forcada, C., Aleácer, E., Garcerá, M.D. and Martinez, R. (1996) Differences in the midgut proteolytic activity of two *Helothis virescens* strains, one susceptible and one resistant to *Bacillus thuringiensis* toxins. *Archives of Insect Biochemistry and Physiology* 31, 257–272.

Ford, M.G., Greenwood, R. and Thomas, P.J. (1981) The kinetics of insecticide action. 1. The properties of a mathematical model describing insect pharmacokinetics. *Pesticide Science* 12, 175–198.

Foster, S.P., Kift, N.B., Baverstock, J., Sime, S., Reynolds, K., Jones, J.E., Thompson, R. and Tatchell, G.M. (2003) Association of MACE-based insecticide resistance in *Myzus persicae* with reproductive rate, response to alarm pheromone and vulnerability to attack by *Aphidius colemani*. *Pest Management Science* 59, 1169–1178.

Fournier, D. (2005) Mutations of acetylcholinesterase which confer insecticide resistance in insect populations. *Chemico-biological Interactions* 157/158, 257–261.

Gahan, L.J., Gould, F. and Heckel, D.G. (2001) Identification of a gene associated with *Bt* resistance in *Heliothis virescens*. *Science* 293, 857–860.

Gardiner, E.M.M. and Plapp, F.W. (1997) Insecticide uptake and decreased uptake resistance in the house fly (*Diptera: Muscidae*): a study with avermectin. *Journal of Economic Entomology* 90, 261–266.

Gerrard, B., Stewart, C. and Dean, M. (1993) Analysis of Mdr50-A *Drosophila* P-glycoprotein/multidrug-resistance gene homolog. *Genomics* 17, 83–88.

Giacobini, E. (2000) *Cholinesterases and Cholinesterase Inhibitors; Basic Preclinical and Clinical Aspects*. Martin Dunitz, London.

Gilbert, L.I., Iatrou, K. and Gill, S.S. (2005) *Comprehensive Insect Molecular Science*, Vols 1–6. Elsevier, Oxford, UK.

Gisselmann, G., Plonka, J., Pusch, H. and Hatt, H. (2004) *Drosophila melanogaster* GRD and LCCH3 subunits form heteromultimeric GABA-gated and cation channels. *British Journal of Pharmacology* 142, 409–413.

Goodman, W.G. and Granger, N.A. (2005) The juvenile hormones. In: Gilbert, L.I., Iatrou, K. and Gill, S.S. (eds) *Comprehensive Insect Molecular Science*, Vol. 3. Elsevier, Oxford, UK, pp. 319–408.

Grant, D.F. and Matsumura, F. (1989) Glutathione *S*-transferase 1 and 2 in susceptible and insecticide resistant *Aedes aegypti*. *Pesticide Biochemistry and Physiology* 33, 132–143.

Greenberg-Levy, S.H., Ishaaya, I., Shaaya, E., Silhacek, D.L. and Oberlander, H. (1995) Hydrolase activity in *Plodia interpunctella*: use of diflubenzuron and *p*-nitroacetanilide as substrates. *Pesticide Biochemistry and Physiology* 52, 157–169.

Greenwood, R., Ford, M.G. and Scarr, A. (2002) Neonicotinoid pharmacokinetics. In: *Proceedings of the 2002 BCPC Conference: Pests and Diseases*, Vol. 1. BCPC Publications, Alton, UK, pp. 153–160.

Greenwood, R., Salt, D. and Ford, M. (2007) Pharmacokinetics: computational versus experimental approaches to optimize insecticidal chemistry In: Ishhaya, I., Nauen, R. and Horowitz, R.I. (eds) *Insecticides Design Using Advanced Technologies*. Springer, Berlin, pp. 41–66.

Griffitts, J.S. and Aroian, R.V. (2005) Many roads to resistance: how invertebrates adapt to *Bt* toxins. *BioEssays* 27, 614–624.

Guerrero, F.D. (2000) Cloning of a horn fly cDNA, Hi E7, encoding an esterase whose transcript concentration is elevated in diazinon-resistant flies. *Insect Biochemistry and Molecular Biology* 30, 1107–1115.

Guerrero, F.D., Jamroz, R.C., Kammlah, D. and Kunz, S.E. (1997) Toxicological and molecular characterization of pyrethroid-resistant horn flies, *Haematobia irritans*: identification of kdr and super-kdr point mutations. *Insect Biochemistry and Molecular Biology* 27, 745–755.

Gunning, R.V. and Devonshire, A.L. (2003) Negative cross-resistance between indoxacarb and pyrethroids in the cotton bollworm, *Helicoverpa armigera*, in Australia: a tool for resistance management. In: *Proceedings of the BCPC International Congress – Crop Science and Technology*, Vol. 2. BCPC Publications, Alton, UK, pp. 789–794.

Harel, M., Kryger, G., Rosenberry, T.L., Mallender, W.D., Lewis, T., Fletcher, R.J., Guss, J.M., Silman, I. and Sussman, J.L. (2000) Three-dimensional structures of *Drosophila melanogaster* acetylcholinesterase and of its complexes with two potent inhibitors. *Protein Science* 9, 1063–1072.

Hartley, C.J., Newcomb, R.D., Russell, R.J., Yong, C.G., Stevens, J.R., Yeates, D.K., La Salle, J. and Oakeshott, J.G. (2006) Amplification of DNA from preserved specimens shows blowflies were preadapted for the rapid evolution of insecticide resistance. *Proceedings of the National Academy of Sciences USA* 103, 8757–8762.

Hayes, J.D., Flanagan, J.U. and Jowsey, I.R. (2005) Glutathione transferases. *Annual Review of Pharmacology and Toxicology* 45, 51–88.

Hedley, D., Khambay, B.P.S., Hooper, A.M., Thomas, R.D, and Devonshire, A.L. (1998) Proinsecticides effective against insecticide-resistant peach-potato aphid (*Myzus persicae* (Sulzer)). *Pesticide Science* 53, 201–208 and 54,188.

Heidari, R., Devonshire, A.L., Campbell, B.E., Bell, K.L., Dorrian, S.J., Oakeshott, J.G. and Russell, R.J. (2004) Hydrolysis of organophosphorus insecticides by *in vitro* modified carboxylesterase E3 from *Lucilia cuprina*. *Insect Biochemistry and Molecular Biology* 34, 353–363.

Heidari, R., Devonshire, A.L., Campbell, B.E., Dorrian, S.J. and Oakeshott, J.G. (2005) Hydrolysis of pyrethroids by carboxylesterases from *Lucilia cuprina* and *Drosophila melanogaster* with active sites modified by *in vitro* mutagenesis. *Insect Biochemistry and Molecular Biology* 35, 597–609.

Hemingway, J. and Ranson, H. (2000) Insecticide resistance in insect vectors of human disease. *Annual Review of Entomology* 45, 371–391.

Hemingway, J., Hawkes, N.J., McCarroll, L. and Ranson, H. (2004) The molecular basis of insecticide resistance in mosquitoes. *Insect Biochemistry and Molecular Biology* 34, 653–665.

Henrick, C.A., Staal, G.B. and Siddall, J.B. (1973) Alkyl 3,7,11-trimethyl-2,4-dodecadienoates, a new class of potent insect growth regulators with juvenile hormone activity. *Journal of Agricultural and Food Chemistry* 21, 354–359.

Henry, R.A. and Byington, K.H. (1976) Inhibition of glutathione-S-aryltransferase from rat liver by organogermanium, lead and tin compounds. *Biochemical Pharmacology* 25, 2291–2295.

Herrero, S., Oppert, B. and Ferré, J. (2001) Different mechanisms of resistance to *Bacillus thuringiensis* toxins in the Indianmeal moth. *Applied and Environmental Microbiology* 67, 1085–1089.

Hille, B. (1977) Local anesthetics: hydrophilic and hydrophobic pathways for the drug–receptor reaction. *Journal of General Physiology* 69, 497–515.

Hollingworth, R.M. (1971) Comparative metabolism and selectivity of organophosphate and carbamate insecticides. *Bulletin of the World Health Organization* 44, 155–170.

Hollingworth, R.M. (2001) Inhibitors and uncouplers of oxidative phosphorylation. In: Krieger, R.I. (ed.) *Handbook of Pesticide Toxicology*, 2nd ed., Vol. 2. Academic Press, San Diego, California, pp. 1169–1262.

Hollingworth, R.M., Kurihara, N., Miyamoto, J., Otto, S. and Paulson, G.D. (1995) Detection and significance of active metabolites of agrochemicals and related xenobiotics in animals. *Pure and Applied Chemistry* 67, 1487–1532.

Hosie, A.M., Baylis, H.A., Buckingham, S.D. and Sattelle, D.B. (1995) Actions of the insecticide fipronil on dieldrin-sensitive and dieldrin-resistant GABA receptors of *Drosophila melanogaster*. *British Journal of Pharmacology* 115, 909–912.

Hoyer, R.F. and Plapp, F.W. Jr (1968) Insecticide resistance in the house fly: identification of a gene that confers resistance to organo tin insecticides and acts as an intensifier of parathion resistance. *Journal of Economic Entomology* 61, 1269–1276.

Ikeda, T., Zhao, X.L., Kono, Y., Yeh, J.Z. and Narahashi, T. (2003) Fipronil modulation of glutamate-induced chloride currents in cockroach thoracic ganglion neurons. *Neurotoxicology* 24, 807–815.

Iovchev, M., Kodrov, P., Wolstenholme, A.J., Pak, W.L. and Semenov, E.P. (2002) Altered drug resistance and recovery from paralysis in *Drosophila melanogaster* with a deficient histamine-gated chloride channel. *Journal of Neurogenetics* 16, 249–261.

Ishaaya, I. (2001) *Biochemical Sites of Insecticide Action and Resistance*. Springer, Heidelberg, Germany.

Jaschke, P. and Nauen, R. (2005) Neonicotinoid insecticides. In: Gilbert, L.I., Iatrou, K. and Gill, S.S. (eds) *Comprehensive Insect Molecular Science*, Vol. 5. Elsevier, Oxford, UK, pp. 53–105.

Kane, N.S., Hirschberg, B., Qian, S., Hunt, D., Thomas, B., Brochu, R., Ludmerer, S.W., Zheng, Y., Smith, M., Arena, J.P., Cohen, C.J., Schmatz, D., Warmke, J. and Cully, D.F. (2000) Drug-resistant *Drosophila* indicate glutamate-gated chloride channels are targets for the antiparasitics nodulisporic acid and ivermectin. *Proceedings of the National Academy of Sciences USA* 97, 13949–13954.

Karlin, A. (2002) Emerging structure of the nicotinic acetylcholine receptors. *Nature Reviews Neuroscience* 3, 102–114.

Kasai, S. and Scott, J.G. (2001) Expression and regulation of CYP6D3 in the house fly, *Musca domestica* (L.). *Insect Biochemistry and Molecular Biology* 32, 1–8.

Kasai, Y., Konno. T. and Dauterman, W.C. (1992) Role of phosphotriesterase hydrolases in the detoxication of organophophorus insecticides. In: Chambers, J.E. and Levi, P.E. (eds) *Organophosphates: Chemistry, Fate and Effects*. Academic Press, San Diego, California, pp. 169–182.

Kerboeuf, D., Blackhall, W., Kaminsky, R. and von Samson-Himmelstjerna, G. (2003) P-glycoprotein in helminths: function and perspectives for anthelmintic treatment and reversal of resistance. *International Journal of Antimicrobial Agents* 22, 332–346.

Ketterman, A.J., Prommeenate, P., Boonchauy, C., Chanama, U., Leetachewa, S., Promtet, N. and Prapanthadara, L. (2001) Single amino acid changes outside the active site significantly affect activity of glutathione *S*-transferases. *Insect Biochemistry and Molecular Biology* 31, 65–74.

Khambay, B.P.S. and Jewess, P.J. (2005) Pyrethroids. In: Gilbert, L.I., Iatrou, K. and Gill, S.S. (eds) *Comprehensive Insect Molecular Science*, Vol. 6. Elsevier, Oxford, UK, pp. 1–29.

Knipple, D.C., Doyle, K.E., Marsella-Herrick, P.A. and Soderlund, D.M. (1994) Tight genetic linkage between the kdr insecticide resistance trait and a voltage-sensitive sodium channel gene in the house fly. *Proceedings of the National Academy of Sciences USA* 91, 2483–2487.

Kobilka, B. (2004) Agonist binding: a multistep process. *Molecular Pharmacology* 65, 1060–1062.

Konno, T., Hodgson, E. and Dauterman, W.C. (1989) Studies on methyl parathion resistance in *Heliothis virescens*. *Pesticide Biochemistry and Physiology* 33, 189–199.

Konno, T., Kasai, Y., Rose, R.L., Hodgson, E. and Dauterman, W.C. (1990) Purification and characterization of a phosphotriester hydrolase from methyl parathion-resistant *Heliothis virescens*. *Pesticide Biochemistry and Physiology* 36, 1–13.

Kostaropoulos, I., Papadopoulos, A.I., Metaxakis, A., Boukouvala, E. and Papadopoulou-Mourkidou, E. (2001) Glutathione *S*-transferase in the defence against pyrethroids in insects. *Insect Biochemistry and Molecular Biology* 31, 313–319.

Lagadic, L., Leicht, W., Ford, M.G., Salt, D.W. and Greenwood, R. (1993) Pharmacokinetics of cyfluthrin in *Spodoptera littoralis* (Boisd.) 1. *In vivo* distribution and elimination of

[¹⁴C]cyfluthrin in susceptible and pyrethroid-resistant larvae. *Pesticide Biochemistry and Physiology* 45, 105–115.

Langton K.P., Henderson P.J.F., and Herbert R.B. (2005) Antibiotic resistance: multidrug efflux proteins, a common transport mechanism? *Natural Product Reports* 22, 439–451.

Lanning, C.L., Ayad, H.M. and Abou-Donia, M.B. (1996a) P-glycoprotein involvement in cuticular penetration of [¹⁴C]thiodicarb in resistant tobacco budworms. *Toxicology Letters* 85, 127–133.

Lanning, C.L., Fine, R.L., Corcoran, J.J., Ayad, H.M., Rose, R.L. and Abou-Donia, M.B. (1996b) Tobacco budworm P-glycoprotein: biochemical characterization and its involvement in pesticide resistance. *Biochimica et Biophysica Acta* 1291, 155–162.

Le Bourg, E. (2001) Oxidative stress, aging and longevity in *Drosophila melanogaster*. *FEBS Letters* 498, 183–186.

Lee, S.H., Smith, T.J., Knipple, D.C. and Soderlund, D.M. (1999) Mutations in the house fly Vssc1 sodium channel gene associated with super-kdr resistance abolish the pyrethroid sensitivity of Vssc1/tipE sodium channels expressed in *Xenopus* oocytes. *Insect Biochemistry and Molecular Biology* 29, 185–194.

Lee, S.J., Tomizawa, M. and Casida, J.E. (2003) Nereistoxin and cartap neurotoxicity attributable to direct block of the insect nicotinic receptor/channel. *Journal of Agricultural and Food Chemistry* 51, 2646–2652.

Le Goff, G., Boundy, S., Daborn, P.J., Yen, J.L., Sofer, L., Lind, R., Sabourault, C., Madi-Ravazzi, L. and Ffrench-Constant, R.H. (2003) Microarray analysis of cytochrome P450 mediated insecticide resistance in *Drosophila*. *Insect Biochemistry and Molecular Biology* 33, 701–708.

Le Goff, G., Hamon, A., Berge, J.B. and Amichot, M. (2005) Resistance to fipronil in *Drosophila simulans*: influence of two point mutations in the Rdl GABA receptor subunit. *Journal of Neurochemistry* 92, 1295–1305.

Lenormand, T., Guillemaud, T., Bourguet, D. and Raymond, M. (1998) Appearance and sweep of a gene duplication: adaptive response and potential for new functions in the mosquito *Culex pipiens*. *Evolution* 52, 1705–1712.

Li, H.R., Oppert, B., Higgins, R.A., Huang, F.N., Zhu, K.Y. and Buschman, L.L. (2004) Comparative analysis of proteinase activities of *Bacillus thuringiensis*-resistant and -susceptible *Ostrinia nubilalis* (Lepidoptera: Crambidae). *Insect Biochemistry and Molecular Biology* 34, 753–762.

Li, H.R., Oppert, B., Higgins, R.A., Huang, F.N., Buschman, L.L., Gao, J.R. and Zhu, K.Y. (2005) Characterization of cDNAs encoding three trypsin-like proteinases and mRNA quantitative analysis in *Bt*-resistant and -susceptible strains of *Ostrinia nubilalis*. *Insect Biochemistry and Molecular Biology* 35, 847–860.

Liapakis, G., Chan, W.C., Papadokostaki, M. and Javitch, J.A. (2004) Synergistic contributions of the functional groups of epinephrine to its affinity and efficacy at the b2 adrenergic receptor. *Molecular Pharmacology* 65, 1181–1190.

Liu, N. and Scott, J.G. (1997) Phenobarbital induction of CYP6D1 is due to a *trans* acting factor on autosome 2 in house flies, *Musca domestica*. *Insect Molecular Biology* 6, 77–81.

Liu, Z., Williamson, M.S., Lansdell, S.J., Denholm, I., Han, Z. and Millar, N.S. (2005) A nicotinic acetylcholine receptor mutation conferring target-site resistance to imidacloprid in *Nilaparvata lugens* (brown planthopper). *Proceedings of the National Academy of Sciences USA* 102, 8420–8425.

Loughney, K., Kreber, R. and Ganetzky, B. (1989) Molecular analysis of the para locus, a sodium channel gene in *Drosophila*. *Cell* 58, 1143–1154.

Ludmerer, S.W., Warren, V.A., Williams, B.S., Zheng, Y.C., Hunt, D.C., Ayer, M.B., Wallace, M.A., Chaudhary, A.G., Egan, M.A., Meinke, P.T., Dean, D.C., Garcia, M.L., Cully, D.F. and Smith, M.M. (2002) Ivermectin and nodulisporic acid receptors in *Drosophila melanogaster* contain both γ-aminobutyric acid-gated Rdl and glutamate-gated GluCl chloride channel subunits. *Biochemistry* 41, 6548–6560.

Lumjuan, N., McCarroll, L., Prapanthadara, L.A., Hemingway, J. and Ranson, H. (2005) Elevated activity of an ε class glutathione transferase confers DDT resistance in the dengue vector, *Aedes aegypti*. *Insect Biochemistry and Molecular Biology* 35, 861–871.

Lund, A.E. and Narahashi, T. (1982) Dose-dependent interaction of the pyrethroid isomers with sodium channels of squid axon membranes. *Neurotoxicology* 3, 11–24.

Martin, T., Ochou, O.G., Vaissayre, M. and Fournier, D. (2003) Oxidases responsible for resistance to pyrethroids sensitize *Helicoverpa armigera* (Hubner) to triazophos in West Africa. *Insect Biochemistry and Molecular Biology* 33, 883–887.

McKenzie, J.A. and O'Farrell, K. (1993) Modification of developmental instability and fitness-malathion-resistance in the Australian sheep blowfly, *Lucilia cuprina*. *Genetica* 89, 67–76.

Menozzi, P., Shi, M.A., Lougarre, A., Tang, Z.H. and Fournier, D. (2004) Mutations of acetylcholinesterase which confer insecticide resistance in *Drosophila melanogaster* populations. BMC *Evolutionary Biology* 4, 4.

Miota, F., Scharf, M.E., Ono, M., Marçon, P., Meinke, L.J., Wright, R.J., Chandler, L.D. and Siegfried, B.D. (1998) Mechanisms of methyl and ethyl parathion resistance in the western corn rootworm (Coleoptera: Chrysomelidae). *Pesticide Biochemistry and Physiology* 61, 39–52.

Miura, K., Oda, M., Makita, S. and Chinzei, Y. (2005) Characterization of the *Drosophila* methoprene-tolerant gene product–juvenile hormone binding and ligand-dependent gene regulation. *FEBS Journal* 272, 1169–1178.

Mohandass, S.M., Arthur, F.H., Zhu, K.Y. and Throne, J.E. (2006) Hydroprene: mode of action, current status in stored-product pest management, insect resistance, and future prospects. *Crop Protection* 25, 902–909.

Morin, S., Biggs, R.W., Sisterson, M.S., Shriver, L., Ellers-Kirk, C., Higginson, D., Holley, D., Gahan, L.J., Heckel, D.G., Carrière, Y., Dennehy, T.J., Brown, J.K. and Tabashnik, B.E. (2003) Three cadherin alleles associated with resistance to *Bacillus thuringiensis* in pink bollworm. *Proceedings of the National Academy of Sciences USA* 100, 5004–5009.

Mota-Sanchez, D., Hollingworth, R.M., Grafius, E.J. and Moyer, D.D. (2006) Resistance and cross-resistance to neonicotinoid insecticides and spinosad in the Colorado potato beetle, *Leptinotarsa decemlineata* (Say) (Coleoptera: Chrysomelidae). *Pest Management Science* 62, 30–37.

Mouchès, C., Pasteur, N., Bergé, J.B., Hyrien, O., Raymond, M., Vincent, B.R., de Silvestri, M. and Georghiou, G.P. (1986) Amplification of an esterase gene is responsible for insecticide resistance in a California *Culex* mosquito. *Science* 233, 778–780.

Mouchès, C., Magnin, M., Bergé, J.B., de Silvestri, M., Beyssat, V., Pasteur, N. and Georghiou, G.P. (1987) Overproduction of detoxifying esterases in organophosphate-resistant *Culex* mosquitoes and their presence in other insects. *Proceedings of the National Academy of Sciences USA* 84, 2113–2116.

Murray, C.L., Quaglia, M., Arnason, J.T. and Morris, C.E. (1994) A putative nicotine pump at the metabolic blood–brain barrier of the tobacco hornworm. *Journal of Neurobiology* 25, 23–34.

Nakaune, R., Adachi, K., Nawata, O., Tomiyama, M., Akutsu, K. and Hibi, T. (1998) A novel ATP-binding cassette transporter involved in multidrug resistance in the phytopathogenic fungus *Penicillium digitatum*. *Applied and Environmental Microbiology* 64, 3983–3988.

Narahashi, T. (1988) Molecular and cellular approaches to neurotoxicology: past, present and future. In: Lunt, G.G. (ed.) *Neurotox '88: Molecular Basis of Drug & Pesticide Action*. Elsevier, New York, pp. 563–582.

Narahashi, T. (2002) Nerve membrane ion channels as the target site of insecticides. *Mini Reviews in Medicinal Chemistry* 2, 419–432.

Nauen, R. and Denholm, I. (2005) Resistance of insect pests to neonicotinoid insecticides: current status and future prospects. *Archives of Insect Biochemistry and Physiology* 58, 200–215.

Nauen, R. and Stumpf, N. (2002) Fluorometric microplate assay to measure glutathione *S*-transferase activity in insects and mites using monochlorobimane. *Analytical Biochemistry* 303, 194–198.

Naumann, K. (1998) Research into fluorinated pyrethroid alcohols – an episode in the history of pyrethroid discovery. *Pesticide Science* 52, 3–20.

Newcomb, R.D., Campbell, P.M., Ollis, D.L., Cheah, E., Russell, R.J. and Oakeshott, J.G. (1997) A single amino acid substitution converts a carboxylesterase to an organophosphorus hydrolase and confers insecticide resistance on a blowfly. *Proceedings of the National Academy of Sciences USA* 94, 7464–7468.

Newcomb, R.D., Gleeson, D.M., Yong, C.G., Russell, R.J. and Oakeshott, J.G. (2005) Multiple mutations and gene duplications conferring organophosphorus insecticide resistance have been selected at the *Rop-1* locus of the sheep blowfly, *Lucilia cuprina*. *Journal of Molecular Evolution* 60, 207–220.

Njue, A.I., Hayashi, J., Kinne, L., Feng, X.P. and Prichard, R.K. (2004) Mutations in the extracellular domains of glutamate-gated chloride channel α 3 and β subunits from ivermectin-resistant *Cooperia oncophora* affect agonist sensitivity. *Journal of Neurochemistry* 89, 1137–1147.

Oakeshott, J.G., Claudianos, C., Russell, R.J. and Robin, G.C. (1999) Carboxyl/cholinesterases: a case study of the evolution of a successful multigene family. *BioEssays* 21, 1031–1042.

Oakeshott, J.G., Horne, I., Sutherland, T.D. and Russell, R.J. (2003) The genomics of insecticide resistance. *Genome Biology* 4, 202.1–202.4.

Oakeshott, J.G., Claudianos, C., Campbell, P.M., Newcomb, R.D. and Russell, R.J. (2005) Biochemical genetics and genomics of insect esterases. In: Gilbert, L.I., Iatrou, K. and Gill, S.S. (eds) *Comprehensive Insect Molecular Science*, Vol. 5. Elsevier, Oxford, UK, pp. 309–381.

Oakley, A.J., Lo Bello, M., Nuccetelli, M., Mazzetti, A.P. and Parker, M.W. (1999) The ligandin (non-substrate) binding site of human Pi class glutathione transferase is located in the electrophile binding site (H-site). *Journal of Molecular Biology* 291, 913–926.

Oakley, A.J., Harnnoi, T., Udomsinprasert, R., Jirajaroenrat, K., Ketterman, A.J. and Wilce, M.C.J. (2001) The crystal structures of glutathione *S*-transferases isozymes 1–3 and 1–4 from *Anopheles dirus* species B. *Protein Science* 10, 2176–2185.

Oi, M. (1996) Application of toxicokinetic approaches to insecticide resistance studies. *Journal of Pesticide Science* 21, 379–388.

Oi, M., Dauterman, W.C. and Motoyama, N. (1990) Biochemical factors responsible for an extremely high level of diazinon resistance in a housefly strain. *Journal of Pesticide Science* 15, 217–224.

Oi, M., Dauterman, W.C. and Motoyama, N. (1993) Toxicokinetic analysis of injected diazinon and diazoxon in resistant and susceptible houseflies, *Musca domestica* L. *Applied Entomology and Zoology* 28, 59–69.

Oppenoorth, F.J. and van Asperen, K. (1960) Allelic genes in the housefly producing modified enzymes that cause organophosphate resistance. *Science* 132, 298–299.

Oppert, B., Kramer, K.J., Beeman, R.W., Johnson, D. and McGaughey, W.H. (1997) Proteinase-mediated insect resistance to *Bacillus thuringiensis* toxins. *Journal of Biological Chemistry* 272, 23473–23476.

O'Reilly, A.Q., Khambay, B.P.S., Williamson, M.S., Field, L.M., Wallace, B.A. and Davies, T.G.E. (2006) Modeling insecticide binding sites at the voltage-gated sodium channel. *Biochemical Journal* 396, 255–263.

Orr, N., Watson, G.B., Hasler, J.M., Mitchell, J.C., Gustafson, G.D., Gifford, J., Cook, K.R., Chouinard, S.W. and Geng, C. (2006) Molecular target site for spinosad: identification and expression of the *Drosophila melanogaster rsn* gene. *Abstracts of the 2006 Entomological Society of America Annual Meeting*, Indianapolis, Indiana, abstr. D0228.

Parkes, T.L., Hilliker, A.J. and Phillips, J.P. (1993) Genetic and biochemical analysis of glutathione-S-transferase in the oxygen defense system of *Drosophila melanogaster*. *Genome* 36, 1007–1014.

Pauron, D., Barhanin, J., Amichot, M., Pralavorio, M., Berge, J.-B. and Lazdunski, M. (1989) Pyrethroid receptor in the insect Na+ channel: alteration of its properties in pyrethroid-resistant flies. *Biochemistry* 28, 1673–1677.

Pimprale, S.S., Besco, C.L., Bryson, P.K. and Brown, T.M. (1997) Increased susceptibility of pyrethroid-resistant tobacco budworm (Lepidoptera: Noctuidae) to chlorfenapyr. *Journal of Economic Entomology* 90, 49–54.

Plapp, F.W. (1976) Biochemical genetics of insecticide resistance. *Annual Review of Entomology* 21, 179–197.

Plapp, F.W. Jr (1984) The genetic basis of insecticide resistance in the house fly: evidence that a single locus plays a major role in metabolic resistance to insecticides. *Pesticide Biochemistry and Physiology* 22, 194–201.

Plapp, F.W. Jr and Hoyer, R.F. (1968) Insecticide resistance in the housefly: decreased rate of absorption as the mechanism of action of a gene that acts as an intensifier of resistance. *Journal of Economic Entomology* 61, 1298–1303.

Pospischil, R., Szomm, K., Londershausen, M., Schroder, I., Turberg, A. and Fuchs, R. (1996) Multiple resistance in the larger house fly *Musca domestica* in Germany. *Pesticide Science* 48, 333–341.

Prasanthi, K., Muralidhara and Rajini, P.S. (2005) Fenvalerate-induced oxidative damage in rat tissues and its attenuation by dietary sesame oil. *Food and Chemical Toxicology* 43, 299–306.

Qiao, D., Seidler, F.J. and Slotkin, T.A. (2005) Oxidative mechanisms contributing to the developmental neurotoxicity of nicotine and chlorpyrifos. *Toxicology and Applied Pharmacology* 206, 17–26.

Ranson, H. and Hemingway, J. (2005) Glutathione transferases. In: Gilbert, L.I., Iatrou, K. and Gill, S.S. (eds) *Comprehensive Insect Molecular Science*, Vol. 5. Elsevier, Oxford, UK, pp. 383–402.

Ratra, G.S., Erkkila, B.E., Weiss, D.S. and Casida, J.E. (2002) Unique insecticide specificity of human homomeric ρ1 GABA(C) receptor. *Toxicology Letters* 129, 47–53.

Raymond, M., Heckel, D.G. and Scott, J.G. (1989) Interactions between pesticide genes: model and experiment. *Genetics* 123, 543–551.

Rossignol, D.P. (1988) Reduction in number of nerve membrane sodium channels in pyrethroid resistant house flies. *Pesticide Biochemistry and Physiology* 32, 146–152.

Rugg, D., Buckingham, S.D., Sattelle, D.B. and Jansson, R.K. (2005) The insecticidal macrocyclic lactones. In: Gilbert, L.I., Iatrou, K. and Gill, S.S. (eds) *Comprehensive Insect Molecular Science*, Vol. 5. Elsevier, Oxford, UK, pp. 25–52.

Russell, R.J., Claudianos, C., Campbell, P.M., Horne, I., Sutherland, T.D. and Oakeshott, J.G. (2004) Two major classes of target site insensitivity mutations confer resistance to organophosphate and carbamate insecticides. *Pesticide Biochemistry and Physiology* 79, 84–93.

Sabourault, C., Guzov, V.M., Koener, J.F., Claudianos, C., Plapp, F.W. Jr. and Feyereisen, R. (2001) Overproduction of a P450 that metabolizes diazinon is linked to a loss-of-function in the chromosome 2 ali-esterase (*MdαE7*) gene in resistant house flies. *Insect Molecular Biology* 10, 609–618.

Salgado, V.L. and Sparks, T.C. (2005) The spinosyns: chemistry, biochemistry, mode of action, and resistance. In: Gilbert, L.I., Iatrou, K. and Gill, S.S. (eds) *Comprehensive Insect*

Molecular Science, Vol. 6. Elsevier, Oxford, UK, pp. 137–173.

Sanchez-Arroyo, H., Koehler, P.G. and Valles, S.M. (2001) Effects of the synergists piperonyl butoxide and *S,S,S*-tributyl phosphorotrithioate on propoxur pharmacokinetics in *Blattella germanica* (Blattodea: Blattellidae). *Journal of Economic Entomology* 94, 1209–1216.

Satoh, T. and Hosokawa, M. (1998) The mammalian carboxylesterases: from molecules to functions. *Annual Review of Pharmacology and Toxicology* 38, 257–288.

Sattelle, D.B., Jones, A.K., Sattelle, B.M., Matsuda, K., Reenan, R. and Biggin, P.C. (2005) Edit, cut and paste in the nicotinic acetylcholine receptor gene family of *Drosophila melanogaster*. *BioEssays* 27, 366–376.

Sawicki, R.M. (1973) Recent advances in the study of the genetics of resistance in the housefly *Musca domestica*. *Pesticide Science* 4, 501–512.

Sawicki, R.M. and Lord, K.A. (1970) Some properties of a mechanism delaying penetration of insecticides into houseflies [*Musca domestica*]. *Pesticide Science* 1, 213–217.

Sayyed, A.H., Raymond, B., Ibiza-Palacios, M.S., Escriche, B. and Wright, D.J. (2004) Genetic and biochemical characterization of field-evolved resistance to *Bacillus thuringiensis* toxin Cry1ac in the diamondback moth, *Plutella xylostella*. *Applied and Environmental Microbiology* 70, 7010–7017.

Sayyed, A.H., Gatsi, R., Ibiza-Palacios, M.S., Escriche, B., Wright, D.J. and Crickmore, N. (2005) Common, but complex, mode of resistance of *Plutella xylostella* to *Bacillus thuringiensis* toxins Cry1ab and Cry1ac. *Applied and Environmental Microbiology* 71, 6863–6869.

Scharf, M.E., Siegfried, B.D., Meinke, L.J. and Chandler, L.D. (2000) Fipronil metabolism, oxidative sulfone formation and toxicity among organophosphate- and carbamate-resistant and susceptible western corn rootworm populations. *Pest Management Science* 56, 757–766.

Schlenke, T.A. and Begun, D.J. (2004) Strong selective sweep associated with a transposon insertion in *Drosophila simulans*. *Proceedings of the National Academy of Sciences USA* 101, 1626–1631.

Scott, J.G. (1999) Cytochromes P450 and insecticide resistance. *Insect Biochemistry and Molecular Biology* 29, 757–777.

Scott, J.G. and Zhang, L. (2003) The house fly aliesterase gene (*MdaE7*) is not associated with insecticide resistance or P450 expression in three strains of house fly. *Insect Biochemistry and Molecular Biology* 33, 139–144.

Sheehan, D., Meade, G., Foley, V.M. and Dowd, C.A. (2001) Structure, function and evolution of glutathione transferases: implications for classification of non-mammalian members of an ancient enzyme superfamily. *Biochemical Journal* 360, 1–16.

Shemshedini, L. and Wilson, T.G. (1990) Resistance to juvenile hormone and an insect growth-regulator in *Drosophila* is associated with an altered cytosolic juvenile hormone-binding protein. *Proceedings of the National Academy of Sciences USA* 87, 2072–2076.

Sheppard, D.C. and Joyce, J.A. (1998) Increased susceptibility of pyrethroid-resistant horn flies (Diptera: Muscidae) to chlorfenapyr. *Journal of Economic Entomology* 91, 398–400.

Shi, M.A., Lougarre, A., Alies, C., Frémaux, I., Tang, Z.H., Stojan, J. and Fournier, D. (2004) Acetylcholinesterase alterations reveal the fitness cost of mutations conferring insecticide resistance. *BMC Evolutionary Biology* 4, 5.

Shono, T., Kasai, S., Kamiya, E., Kono, Y. and Scott, J.G. (2002) Genetics and mechanisms of permethrin resistance in the YPER strain of house fly. *Pesticide Biochemistry and Physiology* 73, 127–136.

Silver, K. and Soderlund, D.M. (2005) State-dependent block of rat $Na_v1.4$ sodium channels expressed in *Xenopus* oocytes by pyrazoline-type insecticides. *Neurotoxicology* 26, 397–406.

Skalsky, H.L. and Guthrie, F.E. (1977) Affinities of parathion, DDT, dieldrin, and carbaryl for macromolecules in the blood of the rat and American cockroach and the competitive interaction of steroids. *Pesticide Biochemistry and Physiology* 7, 289–296.

Soderlund, D.M. (1997) Molecular mechanisms of insecticide resistance. In: Sjut, V. and Butters, J.A. (eds) *Molecular Mechanisms of Resistance to Agrochemicals*. Springer, Berlin, pp. 21–56.

Soderlund, D.L. (2005) Sodium channels. In: Gilbert, L.I., Iatrou, K. and Gill, S.S. (eds) *Comprehensive Insect Molecular Science*, Vol. 5. Elsevier, Oxford, UK, pp. 1–24.

Soderlund, D.M. and Bloomquist, J.R. (1990) Molecular mechanisms of insecticide resistance. In: Roush, R.T. and Tabashnik, B.E. (eds) *Pesticide Resistance in Arthropods*. Chapman and Hall, New York, pp. 58–96.

Soderlund, D.M. and Knipple, D.C. (2003) The molecular biology of knockdown resistance to pyrethroid insecticides. *Insect Biochemistry and Molecular Biology* 33, 563–577.

Soderlund, D.L., Ingles, P.J., Lee, S.H., Smith, T.J. and Knipple, D.C. (2002) The molecular mechanism of knockdown resistance. In: Clark, J.M. and Yamaguchi, I. (eds) *Agrochemical Resistance: Extent, Mechanism, and Detection*. ACS Symposium Series No. 808. American Chemical Society, Washington, DC, pp. 77–89.

Song, W., Liu, Z. and Dong, K. (2006) Molecular basis of differential sensitivity of insect sodium channels to DCJW, a bioactive metabolite of the oxadiazine insecticide indoxacarb. *Neurotoxicology* 27, 237–244.

Srinivas, R., Udikeri, S.S., Jayalakshmi, S.K. and Sreeramulu, K. (2004) Identification of factors responsible for insecticide resistance in *Helicoverpa armigera*. *Comparative Biochemistry and Physiology. Toxicology & Pharmacology* 137, 261–269.

Sussman, J.L., Harel, M., Frolow, F., Oefner, C., Goldman, A., Toker, L. and Silman, I. (1991) Atomic structure of acetylcholinesterase from *Torpedo californica*: a prototypic acetylcholine-binding protein. *Science* 253, 872–879.

Szeicz, F.M., Plapp, F.W. and Vinson, S.B. (1973) Tobacco budworm: penetration of several insecticides into the larvae. *Journal of Economic Entomology* 66, 9–15.

Tan, J., Liu, Z., Tsai, T.D., Valles, S.M., Goldin, A.L. and Dong, K. (2002) Novel sodium channel gene mutations in *Blattella germanica* reduce the sensitivity of expressed channels to deltamethrin. *Insect Biochemistry and Molecular Biology* 32, 445–454.

Tan, J., Liu, Z., Wang, R., Huang, Z.Y., Chen, A.C., Gurevitz, M. and Dong, K. (2005) Identification of amino acid residues in the insect sodium channel critical for pyrethroid binding. *Molecular Pharmacology* 67, 513–522.

Taniai, K., Inceoglu, A.B., Kenji Yukuhiro, K. and Hammock, B.D. (2003) Characterization and cDNA cloning of a clofibrate-inducible microsomal epoxide hydrolase in *Drosophila melanogaster*. *European Journal of Biochemistry* 270, 4696–4705.

Taylor, M.F.J., Heckel, D.G., Brown, T.M., Kreitman, M.E. and Black, B. (1993) Linkage of pyrethroid insecticide resistance to a sodium channel locus in the tobacco budworm. *Insect Biochemistry and Molecular Biology* 23, 763–775.

Trainer, V.L., McPhee, J.C., Boutelet-Bochan, H., Baker, C., Scheuer, T., Babin, D., Demoute, J.-P., Guedin, D. and Catterall, W.A. (1997) High affinity binding of pyrethroids to the α-subunit of brain sodium channel. *Molecular Pharmacology* 51, 651–657.

Udomsinprasert, R. and Ketterman, A.J. (2002) Expression and characterization of a novel class of glutathione *S*-transferase from *Anopheles dirus*. *Insect Biochemistry and Molecular Biology* 32, 425–433.

Udomsinprasert, R., Pongjaroenkit, S., Wongsantichon, J., Oakley, A.J., Prapanthadara, L., Wilce, M.C.J. and Ketterman, A.J. (2005) Identification, characterization and structure of a new δ class glutathione transferase isoenzyme. *Biochemical Journal* 388, 763–771.

Vais, H., Williamson, M.S., Goodson, S.J., Devonshire, A.L., Warmke, J.W., Usherwood, P.N.R. and Cohen, C. (2000) Activation of *Drosophila* sodium channels promotes modification by deltamethrin: reductions in affinity caused by knock-down resistance mutations. *Journal of General Physiology* 115, 305–318.

Vais, H., Williamson, M.S., Devonshire, A.L. and Usherwood, P.N. (2001) The molecular interactions of pyrethroid insecticides with insect and mammalian sodium channels. *Pest Management Science* 57, 877–888.

Vais, H., Atkinson, S., Pluteanu, F., Goodson, S.J., Devonshire, A.L., Williamson, M.S. and Usherwood, P.N. (2003) Mutations of the para sodium channel of *Drosophila melanogaster* identify putative binding sites for pyrethroids. *Molecular Pharmacology* 64, 914–922.

Vijverberg, H.P.M., van der Zalm, J.M. and van den Berchen, J. (1982) Similar mode of action of pyrethroids and DDT on sodium channel gating in myelinated nerves. *Nature* 295, 601–603.

Vilanova, E. and Sogorb, M.A. (1999) The role of phosphotriesterases in the detoxication of organophosphorus compounds. *Critical Reviews in Toxicology* 29, 21–57.

Vontas, J.G., Small, G.J. and Hemingway, J. (2001) Glutathione *S*-transferases as antioxidant defence agents confer pyrethroid resistance in *Nilaparvata lugens*. *Biochemical Journal* 357, 65–72.

Vontas, J.G., Small, G.J., Nikou, D.C., Ranson, H. and Hemingway, J. (2002) Purification, molecular cloning and heterologous expression of a glutathione *S*-transferase involved in insecticide resistance from the rice brown planthopper, *Nilaparvata lugens*. *Biochemical Journal* 362, 329–337.

Warmke, J.W., Reenan, R.A.G., Wang, P., Qian, S., Arena, J.P., Wang, J., Wunderler, D., Liu, K., Kaczorowski, G.J., Van der Ploeg, L.H.T., Ganetzky, B. and Cohen, C.J. (1997) Functional expression of *Drosophila* para sodium channels: modulation by the membrane protein tipE and toxin pharmacology. *Journal of General Physiology* 110, 119–133.

Wei, S.H., Clark, A.G. and Syvanen, M. (2001) Identification and cloning of a key insecticide-metabolizing glutathione *S*-transferase (MdGST-6A) from a hyper insecticide-resistant strain of the housefly *Musca domestica*. *Insect Biochemistry and Molecular Biology* 31, 1145–1153.

Weill, M., Fort, P., Berthomieu, A., Dubois, M.P., Pasteur, N. and Raymond, M. (2002) A novel acetylcholinesterase gene in mosquitoes codes for the insecticide target and is non-homologous to the *ace* gene in *Drosophila*. *Proceedings of the Royal Society of London. Series B, Biological Sciences* 269, 2007–2016.

Weill, M., Lutfalla, G., Mogensen, K., Chandre, F., Berthomieu, A., Berticat, C., Pasteur, N., Philips, A., Fort, P. and Raymond, M. (2003) Insecticide resistance in mosquito vectors. *Nature* 423, 136–137 and 425, 366.

Welling, W. (1977) Dynamic aspects of insect–insecticide interactions. *Annual Review of Entomology* 22, 53–78.

Welling, W. (1979) Toxicodynamics of insecticidal action – introduction. *Pesticide Science* 10, 540–546.

Welling, W., DeVries, J.W., Paterson, G.D. and Duffy, M.R. (1983) Toxicodynamics of malaoxon in house-flies. *Pesticide Biochemistry and Physiology* 20, 360–372.

Wheelock, C.E., Shan, G, and Ottea, J. (2005) Overview of carboxylesterases and their role in the metabolism of insecticides. *Journal of Pesticide Science* 30, 75–83.

Williamson, M.S., Denholm, I., Bell, C.A. and Devonshire, A.L. (1993) Knockdown resistance (kdr) to DDT and pyrethroid insecticides maps to a sodium channel gene locus in the housefly (*Musca domestica*). *Molecular and General Genetics* 240, 17–22.

Williamson, M.S., Martinez-Torres, D., Hick, C.A. and Devonshire, A.L. (1996) Identification of mutations in the housefly para-type sodium channel gene associated with knockdown resistance (kdr) to pyrethroid insecticides. *Molecular and General Genetics* 252, 51–60.

Wilson, T.G. (1993) Involvement of transposable genetic elements in initiating insecticide resistance. *Journal of Economic Entomology* 86, 645–651.

Wilson, T.G. (2001) Resistance of *Drosophila* to toxins. *Annual Review of Entomology* 46, 545–571.

Wilson, T.G. and Ashok, M. (1998) Insecticide resistance resulting from an absence of target-site gene product. *Proceedings of the National Academy of Sciences USA* 95, 14040–14044.

Wilson, T.G., Wang, S., Be o, M. and Farkaš, R. (2006) Wide mutational spectrum of a gene involved in hormone action and insecticide resistance in *Drosophila melanogaster*. *Molecular Genetics and Genomics* 276, 294–303.

Wing, K.D., Schnee, M.E., Sacher, M., and Connair, M. (1998) A novel oxadiazine insecticide is bioactivated in lepidopteran larvae. *Archives of Insect Biochemistry and Physiology* 37, 91–103.

Wing, K.D., Andalaro, J.T., McCann, S.F. and Salgado, V.L. (2005) Indoxacarb and the sodium channel blocker insecticides: chemistry, physiology, and biology in insects. In: Gilbert, L.I., Iatrou, K. and Gill, S.S. (eds) *Comprehensive Insect Molecular Science*, Vol. 6. Elsevier, Oxford, UK, pp. 31–53.

Wongsantichon, J., Harnnoi, T. and Ketterman, A.J. (2003) A sensitive core region in the structure of glutathione *S*-transferases. *Biochemical Journal* 373, 759–765.

Wongtrakul, J., Udomsinprasert, R. and Ketterman, A.J. (2003) Non-active site residues Cys69 and Asp150 affected the enzymatic properties of glutathione *S*-transferase AdGSTD3–3. *Insect Biochemistry and Molecular Biology* 33, 971–979.

Xie, R., Zhuang, M., Ross, L.S., Gomez, I., Oltean, D.I., Bravo, A., Soberon, M. and Gill, S.S. (2005) Single amino acid mutations in the cadherin receptor from *Heliothis virescens* affect its toxin binding ability to Cry1a toxins. *Journal of Biological Chemistry* 280, 8416–8425.

Xu, M., Molento, M., Blackhal, W.I., Ribeiro, P., Beech, R. and Prichard, R. (1998) Ivermectin resistance in nematodes may be caused by alteration of P-glycoprotein homolog. *Molecular and Biochemical Parasitology* 91, 327–335.

Xu, X., Yu, L. and Wu, Y. (2005) Disruption of a cadherin gene associated with resistance to Cry1Ac δ-endotoxin of *Bacillus thuringiensis* in *Helicoverpa armigera*. *Applied and Environmental Microbiology* 71, 948–954.

Young, S.J., Gunning, R.V. and Moores, G.D. (2005) The effect of piperonyl butoxide on pyrethroid-resistance-associated esterases in *Helicoverpa armigera* (Hübner) (Lepidoptera: Noctuidae). *Pest Management Science* 61, 397–401.

Yu, S.J. (1996) Insect glutathione *S*-transferases. *Zoological Studies* 35, 9–19.

Yu, S.J. (2002) Substrate specificity of glutathione *S*-transferases from the fall armyworm. *Pesticide Biochemistry and Physiology* 74, 41–51.

Yu, S.J. and Nguyen, S.N. (1998) Purification and characterization of carboxylamidase from the fall armyworm, *Spodoptera frugiperda* (J.E. Smith). *Pesticide Biochemistry and Physiology* 60, 49–58.

Zhao, X.L., Yeh, J.Z., Salgado, V.L. and Narahashi, T. (2004) Fipronil is a potent open channel blocker of glutamate-activated chloride channels in cockroach neurons. *Journal of Pharmacology and Experimental Therapeutics* 310, 192–201.

Zhu, Y.C., Dowdy, A.K. and Baker, J.E. (1999a) Detection of single-base substitution in an esterase gene and its linkage to malathion resistance in the parasitoid *Anisopteromalus calandrae* (Hymenoptera: Pteromalidae). *Pesticide Science* 55, 398–404.

Zhu, Y.C., Dowdy, A.K. and Baker, J.E. (1999b) Differential mRNA expression levels and gene sequences of a putative carboxylesterase-like enzyme from two strains of the parasitoid *Anisopteromalus calandrae* (Hymenoptera: Pteromalidae). *Insect Biochemistry and Molecular Biology* 29, 417–425.

Zhu, Y.C., Snodgrass, G.L. and Chen, M.S. (2004) Enhanced esterase gene expression and activity in a malathion-resistant strain of the tarnished plant bug, *Lygus lineolaris*. *Insect Biochemistry and Molecular Biology* 34, 1175–1186.

4 Assessing the Risk of the Evolution of Resistance to Pesticides Using Spatially Complex Simulation Models

M.A. Caprio[1], N.P. Storer[2], M.S. Sisterson[3], S.L. Peck[4] and A.H.N. Maia[5]

[1]Department of Entomology and Plant Pathology, Mississippi State University, Mississippi State, Mississippi, USA; [2]Dow AgroSciences LLC, Indianapolis, Indiana, USA; [3]San Joaquin Valley Agricultural Sciences Center, Parlier, California, USA; [4]Department of Biology, Brigham Young University, Provo, Utah, USA; [5]Embrapa Meio Ambiente, Jaguariuna, SP, Brazil

4.1 Introduction

As a class, insects have an extraordinary ability to adapt to diverse environments (Chapman, 1982). At the species level, the life system characteristics typically shared among agricultural pests – the ability to rapidly colonize new or unstable habitats through rapid reproduction and a high motility – include an ability to respond to changing environments (Kennedy and Storer, 2000). Insecticides, presented through either transgenic crops or conventional means, represent a severe environmental challenge in otherwise suitable habitats and pose an interesting evolutionary problem to pest species. Individuals with a genetically based ability to overcome these challenges have higher fitness in more environments than other individuals in the population, and their genotypes should increase in frequency, leading to the evolution of resistance to the toxin. After the first documentation of insect resistance to a synthetic insecticide in 1947 (Metcalf, 1973), the process of resistance evolution quickly drew the interest of population geneticists (Crow, 1957). A desire to proactively manage resistance propelled development of an understanding of the underlying population genetics of resistance evolution. The basic principles of modern resistance management programmes were developed in a series of papers by Taylor and Georghiou (Georghiou and Taylor, 1977a,b; see also Taylor and Georghiou, 1979, 1982, 1983; Taylor, 1986). These papers clearly identified the driving roles of functional dominance of resistance, including the effects of pesticide dose and the problems posed by pesticide decay, and the rate of immigration of susceptible insects in determining rates of adaptation. These concepts were further developed by Tabashnik and Croft (1982, 1985) and others (Curtis *et al.*, 1978; Plapp *et al.*, 1979), who demonstrated that consideration of multiple habitats (treated and untreated fields) is important in understanding the evolution of resistance. They used a continent–island concept in which there is an influx of susceptible alleles but negligible migration of resistance alleles into the untreated population. This concept essentially envisages an infinite supply of susceptible immigrants, whereby the movement of any resistant individuals out of the area of selection would have an insignificant impact on gene frequencies in the surrounding habitats. Comins (1977a,b, 1979) developed a two-patch model that relaxed the assumption of outward migration and allowed the diffusion of resistance alleles into untreated habitats. He concluded that density-dependent processes in population dynamics in each patch are critical components of any multi-patch model and therefore that infinite population models have limited ability in predicting resistance evolution. This idea was further explored by Gould *et al.* (1991), Onstad *et al.* (2001), Storer *et al.* (2003b), Sisterson *et al.* (2004) and Crowder and Onstad (2005), for example.

The development of spatially explicit finite population models (Caprio and Tabashnik, 1992; Mallet and Porter, 1992; Caprio, 1994; Peck *et al.*, 1999; Sisterson *et al.*, 2004) suggested that moderate rates of gene flow, interacting with genetic drift, could actually cause faster resistance evolution than either high or low rates of gene flow under some conditions. Spatial effects and heterogeneous distributions of treated and untreated patches could also impact resistance evolution rates (Peck *et al.*, 1999; Storer, 2003; Sisterson *et al.*, 2005). For the first time, these spatially explicit stochastic models produced estimates of variation due to demographic stochasticity.

From simulation studies using the early models, the conclusion that pesticide dose is a key determinant of rates of resistance evolution highlighted a restriction to devising effective resistance management strategies. While there is the potential for immigration to delay resistance evolution when high doses of an insecticide are used, this strategy was unlikely to be successful with conventional insecticides, because the uneven application and continuous decay of those compounds would expose insects to more moderate doses (Taylor and Georghiou, 1982), altering the dominance of resistance. Host plant resistance factors that are expressed at high levels throughout a crop plant are able to overcome this restriction and open up new avenues for resistance management (Gould, 1986). The advent of transgenic crops that constitutively produced insecticidal proteins accelerated the need and ability to devise effective management strategies (McGaughey and Whalon, 1992; Roush, 1994; Wearing and Hokkanen, 1995; Gould, 1998). Gould (1994) and Roush (1994) argued that combining a consistent high dose in the genetically modified plant with planting of plants that do not express the toxin (a refuge from selection) could delay the evolution of resistance dramatically. The concept that the plants would continue to express the toxin in sufficient quantities over the course of the season to avoid large changes in functional dominance was key to this proposal (Onstad and Gould, 1998). Initially, mixtures of transgenic and non-transgenic seeds seemed like an ideal method to implement this strategy (Tabashnik, 1994a), but

Mallet and Porter (1992) and Davis and Onstad (2000) showed that larval movement among these plant types could shift the dominance of resistance much as Taylor and Georghiou (1982) had described for decaying pesticides. Therefore, the planting of structured refuges was required when insecticidal transgenic crops were introduced in the USA.

The evolution of resistance is a population genetics phenomenon, affected by complex interactions among pest biology and ecology, properties of the pesticide, and pesticide use patterns (Georghiou and Taylor, 1977a,b); and models have always played an important role in our understanding (Taylor, 1983; Tabashnik, 1986). However, in a broader sense, all humans construct models. Indeed, it has been suggested that human knowledge can be seen as the construction of models to understand our social, physical and biological surroundings (Richmond, 2001). Consider the simple everyday example of the decision to make a cup of coffee or tea in the morning. One might evaluate the time and difficulty involved, the cost of the beverage, how much better it tastes than the coffee or tea at work, how much work is waiting at work, etc. Normally, one would not formalize all these parameters, but they have probably been considered, at least fleetingly, by many at one time or another. Similarly, those who work in agricultural systems have conceptual models of what happens in those systems and what inputs or parameters are most important, and, in most cases, have not formalized values for those parameters. While there are many different goals of modelling, one goal is to formalize these conceptual models, to explicitly state the rules and relationships between the components of the models. The model allows one to organize all available data into a coherent framework with clearly stated rules regarding transformation of the model from one state to the next. Once constructed, these models allow one to test hypotheses about effects of changes in parameters or how different components interact and result in observed system behaviour. Models based upon mechanistic processes become experimental systems in which it is possible to develop hypotheses about how one expects the system will behave and then test the system response (Peck, 2004). In many cases, the system is complex enough that it can behave in unexpected ways, and the conclusions are not necessarily an obvious result of the assumptions and rules used in model construction. Indeed, it is often when the model behaves differently from expectations that it is providing the most important information. Such unexpected results may indicate that something is missing from the model, they may guide future research by identifying research needs, or they may indicate that our conceptual understanding of the system is incorrect and requires modification. For example, one early assumption of the high-dose/refuge strategy was that random mating between adults produced in refuges and transgenic fields would maximize delay of resistance evolution (Tabashnik, 1994b). Using a spatially complex model, Caprio (2001) suggested that because of source–sink dynamics, some degree of isolation between a refuge and a highly toxic crop might actually delay resistance much longer than having random mating between the different habitats. Indeed, these predictions were supported by the model of Sisterson *et al.* (2005), and subsequent field work demonstrated that model predictions were consistent with the patterns observed in the field (Carrière *et al.*, 2002; Caprio *et al.*, 2004). As noted by

Alstad and Andow (1995), these dynamics will also create a halo of increased damage surrounding areas planted to refuges. In spatially complex models, Peck *et al.* (1999) and Sisterson *et al.* (2005) found that resistance first evolved in areas with a locally high density of transgenic fields and then spread outwards. Selection in these small areas led to foci of locally higher resistance allele frequencies, an increased frequency of resistant homozygotes and a more rapid rate of resistance evolution. Once resistance became common in these areas, migration spread resistance alleles across the region. The spatial characteristics of the system and the dispersal characteristics of the insect are critical to determining the potential for these spatial dynamics effects (Sisterson *et al.*, 2005).

4.2 Simple Versus Complex Models

As we have seen, there is a wide range in complexity of resistance models, from elegantly simple analytical models to increasingly complex simulations of specific agroecological systems. The simplest models include only a few parameters: (i) initial resistance allele frequency; (ii) strength of selection (e.g. percentage of the population exposed to the insecticidal transgenic crop); and (iii) the dominance of resistance. Because such models have only a few parameters, it is possible to conduct a complete analysis, which provides an understanding of the role of each parameter. It is often possible to construct analytical solutions to identify optimal strategies (e.g. Lenormand and Raymond, 1998). From these simple models, we have developed an understanding of the general principles of resistance evolution from which we can design effective resistance management strategies, such as the potential power of the high-dose/refuge strategy (Gould, 1998). For example, the non-random mating model developed by Caprio (1998), a simple deterministic model with a few parameters, was used for designing preliminary proactive resistance management strategies for *Bt* cotton in the mid-west region of Brazil (Fitt *et al.*, 2006).

However, these models do not inform us about the effectiveness of a specific strategy in a specific system. While simple models provide an understanding of the few parameters that are used, it is unclear how dependent the results are on many of the simplifying assumptions. These models are highly abstracted and simplified versions of field conditions, and the input parameters may actually be combinations of many other parameters that can be measured empirically. Because of this simplification/abstraction process, there may be many underlying processes not explicitly represented. Therefore, it is not possible to evaluate properly the influence of those hidden parameters on the resistance model end points. Of course, even complex models are also simplified/abstracted versions of reality and suffer the same problem, presumably to a lesser extent.

At the opposite end of the spectrum are complex models that may represent several interrelated complex process and incorporate tens, if not hundreds, of parameters. These parameters tend to be less abstract and closer to parameters for which empirically measured estimates exist, although data gaps in each model certainly exist. They tend to be more highly mechanistic simulations of the real world. Complex models are often closed systems with finite populations, incorporate stochastic rather than deterministic processes, add spatial and temporal variability and allow for environmental fluctuations. Such models tend to be much more system-specific, dealing with a particular pest, crop or geography combination (e.g. Peck *et al.*, 1999; Guse *et al.*, 2002; Storer *et al.*, 2003a; Sisterson *et al.*, 2004). While these models have not radically altered our understanding of resistance evolution – the same parameters and the same resistance management strategies have been supported as were supported by simpler, analytical models – they permit insights into the actual risk of resistance in any given real-world scenario, as well as into the potential effectiveness and limitations of a given management tactic.

As an example, landscape ecology is increasingly being integrated into resistance studies. Spatially rich models allow one to incorporate insect movement among multiple habitats to examine issues such as interactions among different crops (Storer *et al.*, 2003a,b). One key component in such models is a shift in focus from a single insect on a single crop to a broader perspective of following an insect as it moves through a variety of habitats during the course of a growing season. These models can include temporal, as well as spatial, variation in such components as crop composition and toxin expression. For example, *Helicoverpa zea* in the Delta region of the mid-southern USA begins the season on wild hosts such as crimson clover and wild geranium (Stadelbacher, 1979; Parker, 2000). Later, it moves on to whorl and eventually ear-stage maize. By mid-July, most maize is no longer suitable habitat for *H. zea*, and, at this time, adults may move into a number of crops, including cotton, sorghum and soybean. Finally, in late summer, some *H. zea* may overwinter in crop fields (where cultivation may yield a significant amount of mortality), while others may move to weedy hosts and overwinter there (Parker 2000; Storer *et al.*, 2003a). Figure 4.1 shows simulated numbers of pupae in each habitat over the course of a season (Parker and Caprio, unpublished data). Population dynamics in each of these habitats, combined with population genetic issues, such as selection by various pesticides and toxins in the different habitats, will affect the impact of pest management decisions. For example, while high adoption rates of transgenic cotton in the Delta may reduce the number of *H. zea* moths emerging from the cotton acreage and reduce overall density in the autumn, density-dependent interactions in maize may allow *H. zea* numbers to build back up to damaging numbers in the early season when maize is a host. Numbers are limited in ear-stage maize, as generally only one adult will emerge from each ear due to cannibalism (Barber, 1936). Thus, density-dependent mortality will be high at this stage when populations are large (which can accelerate resistance evolution) (Storer *et al.*, 2003a), but if population levels decline because of mortality due to transgenic crops, density-dependent mortality may have less impact on the evolution of resistance. When density levels are high, each ear of maize will probably be infested with multiple larvae. *H. zea* larvae are cannibalistic, and the largest larvae are most likely to survive. When a *Bt* toxin is present in kernels, resistant larvae will experience less growth retardation than susceptible larvae and will have a greater chance of surviving not only the *Bt* toxin, but also the density-dependent mortality. In contrast, when densities are low, individual ears will likely be uninfested or have a single larva, and the rate of selection will

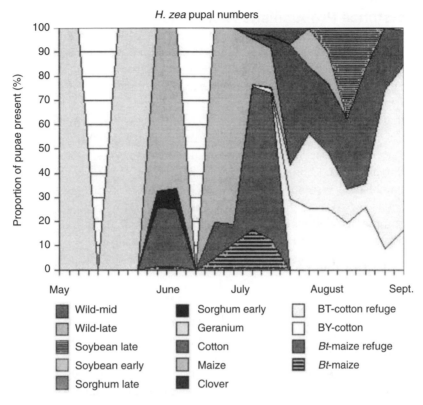

Fig. 4.1. Temporal dynamics of *Helicoverpa zea* pupal numbers in a simulated mid-southern US agroecosystem. The y-axis is the proportion of all pupae present on a given date. (Caprio and Parker, unpublished data.)

only be related to the chance of surviving the *Bt* toxin. These dynamics demonstrate that the two systems interact in a complex manner, and both systems must be understood before one can begin to assess the risk of the evolution of resistance. Current transgenic maize varieties express Cry1Ab or Cry1F insecticidal proteins from *Bacillus thuringiensis* (*Bt*) to control Lepidoptera; while Cry1Ac and combinations of Cry1Ac with Cry1F or Cry2Ab are expressed in current transgenic cotton varieties. Several studies have demonstrated the potential for some level of cross-resistance among these toxins (e.g. Gould *et al.*, 1995a; Jurat-Fuentes *et al.*, 2000; Siqueira *et al.*, 2004). The overall rate of the evolution of resistance to the Cry proteins in cotton will be impacted by pest management decisions in maize (ILSI/HESI, 1999; Storer *et al.*, 2003a). Similarly, altering the planting date of soybeans (or the maturity group planted) can affect the infestation of these habitats which act as refuges to transgenic cotton (Gustafson *et al.*, 2006). Clearly the interactions among the temporal dynamics of the different habitats in a system may also play an important role in the evolution of resistance. The development of exhaustive models to describe such complex systems allows the modeller to consider the effects of these interactions on resistance evolution and potential management approaches.

4.3 Resistance Probabilistic Risk Assessment

The possibility of resistance evolution should be regarded as a hazard associated with the use of any pesticidal technology, similar to the possibility of water, soil or food contamination. The challenge when deploying these technologies responsibly is not just to determine that such hazards exist, but to quantify the risk; that is, quantify the probability of an adverse occurrence and the magnitude of the effect of that occurrence. Where an unacceptable risk is identified, risk mitigation measures (such as spray drift reduction or pre-harvest intervals) can be implemented. Leonard (2000) describes the application of the risk assessment paradigm to resistance management for pesticides in Europe and a set of guidelines for performing the risk assessment. Beyond such qualitative approaches, complex models can be used to determine the magnitude of the risk quantitatively and to design appropriate mitigation measures.

Modelling uncertainty

Uncertainty is inherent in all risk assessments in which simulation models are used to extrapolate information beyond the domain of direct observation (Hoffman and Kaplan, 1999). Different types of simulation models will produce a variety of different results, and interpretation of these results will depend on the underlying structure of the models and the assumptions (explicit and implicit) made. For complex models, interpretation becomes far more challenging as multiple processes are occurring simultaneously, and there is a temptation to use the models in a predictive manner. In quantitatively assessing the risk of the evolution of resistance, information on mean values of selected end points (such as time until resistance allele frequency exceeds some level over some proportion of the *Bt* crop area) is not sufficient. In some cases (skewed or multimodal output distributions) even additional summary measures, like variance or standard error, do not allow risk quantification. To be useful, the model output information should be represented by Monte Carlo probability distributions, used for calculating the probability of occurrence of adverse effects, generally related to the tails of output distributions (see Fig. 4.2). For a given set of input parameters, deterministic models will always produce the same point estimate, therefore not being useful for estimating risks. From a broader perspective, when assessing risk we are most often interested in the tails of probability distributions, in particular the tail that represents the proportion of events that exceed our acceptable criteria. The mean of the distribution, which has been the focus of resistance modelling since its inception, may be relatively uninformative, especially when the shapes of the Monte Carlo distributions of alternative risk scenarios are different. It is conceivable that two strategies with similar mean times to resistance could actually have different risk assessments, because they respond differently to conditions that are responsible for

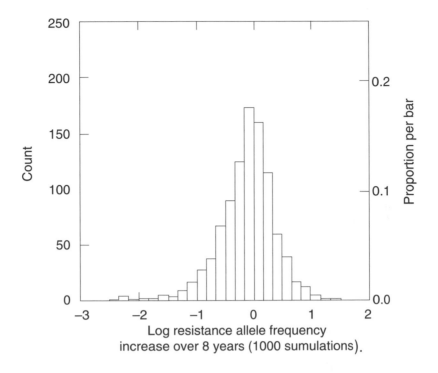

Fig. 4.2. Monte Carlo probability distribution of the mean relative log rate of increase for a resistance allele to a high-dose transgenic event with fitness costs over the initial 8 years of the simulations.

the tails of the distribution, such as rare draws of relatively high values of dominance.

Generally, if one wants to know how sensitive these results are to isolate or conjoint perturbations in the initial input parameters, one runs the model for a series of parameter values without assigning probabilities to those values, the so-called sensitivity analysis. The aim of sensitivity analysis is not producing risk estimates, but evaluating the relative influence of input parameters on the model end points. It is useful for indicating that information on some key parameters should be refined, while, for other ones, improving information would be irrelevant (Isukapalli and Georgopoulus, 2001). An approach widely used for resistance risk assessment is based on choosing a series of likely or unlikely scenarios (best-case, worst-case, etc.) by setting input parameters to some user-defined values and running deterministic models. Point estimates (central tendency measures, like mean) are then produced for each output variable without any associated uncertainty. From a risk assessment perspective, the weakness of this approach (deterministic scenario analysis) is that no probabilities are assigned to the different scenarios. Some scenarios may be seen as unlikely, but there is little to guide the interpretation of the actual probability of occurrence relative to other scenarios. In fact, the relative weights given to the different scenarios are entirely subjective and vary considerably among users of the information. Despite that, the outcomes of model runs with such settings are meaningless for understanding the real world unless there is some way to assess the probability of such a worst case occurring. These results can be easily misinterpreted as being appropriately conservative when designing a resistance management plan or setting policy.

To be useful for risk assessments, models need to produce probabilistic outputs represented by distributions for each end point of interest. These distributions are used to quantify risks generally related to the occurrence of extreme events. Because they are increasingly based on mechanistic processes reflecting real-world biology, complex models lend themselves to be used as predictive tools, just as climate models are used to predict future states of climate, and epidemiological models are used to predict the spread of new diseases. However, none of these models should be used to provide exact predictions, since there is considerable uncertainty inherent to their underlying process. To be used as predictive tools, it is imperative that such models be analysed for their inherent uncertainty, in the same way that weather forecasts incorporate uncertainty by stating there is a certain chance of rain and providing a range of likely amounts, represented by the rain output distribution. For complex models that incorporate stochastic variability and parameter uncertainty, both sources of variability are important in performing risk assessments for the evolution of resistance. We see the formalization of methods for probabilistic risk assessments for transgenic crops using uncertainty analysis as an important new direction in the modelling of resistance evolution. If we are to progress beyond the conventional risk assessment of transgenic crops in which a series of scenarios (best-case, worst-case, etc.) are simulated using deterministic models with no formal guidance on the likelihood of those scenarios, then we must seek to formalize methods for accounting for parameter uncertainty, as well as process stochasticity. These methods allow production of model outputs represented by probability distributions, instead of simple point estimates. Those distributions can then be used for assessing risk, which, by its own definition, should incorporate the probabilistic nature of adverse event predictions. Below we present parameter re-sampling procedures based on Monte Carlo methods that, when combined with stochastic resistance models, will provide an alternative to deterministic scenario analysis.

Uncertainty in model predictions arises from several sources. As insect resistance simulation models seek to simplify highly complex ecological systems in highly variable environments, it cannot be known that the model structure sufficiently, or accurately, describes the system being modelled. The variability inherent to the systems being modelled translates to uncertainty in the appropriate values for input variables and parameters (Fig. 4.3). The values for the input parameters are not perfectly known, and random (or unpredictable) variability in the biotic and abiotic environment (temporal and spatial) causes variability in the value of the parameters. Finally, for a stochastic model of a closed system, different outcomes can result from the same inputs (Fig. 4.4a). The modeller is therefore challenged to account for and describe the uncertainty and variability.

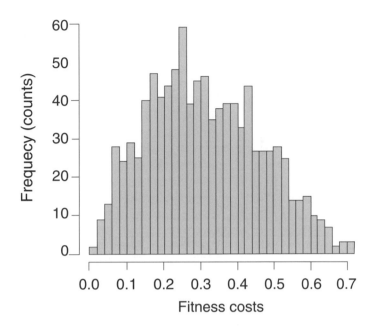

Fig. 4.3. Example of a Monte Carlo probability distribution for the input parameter, Vip resistance allele fitness costs. One thousand draws were made from a beta distribution. The minimum value was 0.0, the maximum value was 0.8, and the most likely value was 0.2. (Caprio and McCaffery, unpublished data.)

Incomplete understanding of the biological processes leads to uncertainty in the true value of a given model parameter. Often modellers need data that have not been considered important from a pest management perspective, and therefore are unavailable. For example, differential mortality of bollworms in refuge patches compared with *Bt* crop patches during winter can have a profound effect on predicted rates of adaptation to *Bt* cotton. However, measuring the level of mortality under natural conditions is extremely difficult to do, and in the past was not regarded as important when the predominant pest management tools were curative rather than prophylactic. To measure winter mortality of bollworms, the field researcher needs to provide good estimates of the final population of pupae in the soil in the autumn and of the population of emerging moths in the spring. Few studies in the literature have provided these estimates, leaving the modeller uncertain as to the correct value to assign to this parameter.

Natural variability in the environment means that the true value of a given parameter varies with space and time. To use the same example, mortality of bollworm pupae in a refuge during winter can depend on the vigour of the

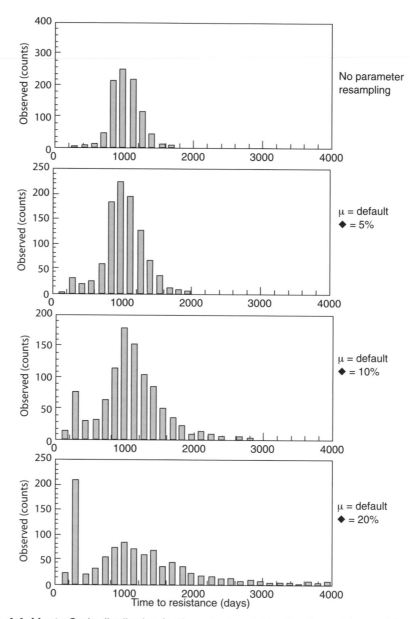

Fig. 4.4. Monte Carlo distribution for the output variable, time to resistance: (a) no parameter re-sampling, uncertainty due to stochastic variability only; (b) mean = default, standard error = 5% of mean; (c) mean = default, standard error = 10% of mean; and (d) mean = default, standard error = 20% of mean.

population as it enters winter (and therefore the quality of the host plant during the preceding larval stage), the weather conditions throughout winter (temperature, moisture), the presence of predators in the soil, the degree of parasitism of the population, the soil type and cultivation activities in the field before moth emergence (perhaps leading to soil compaction, or exposure of pupae to the elements). While the proportion of pupae that survive will clearly be variable, both spatially within and among fields and temporally from year to year, this complex of factors is typically modelled as a single parameter, such as winter survival (Peck *et al.*, 1999; Onstad *et al.*, 2001; Storer 2003; Storer *et al.*, 2003a,b; Sisterson *et al.*, 2004).

Analysing model stochasticity

The use of stochastic models presents an important component of modelling uncertainty, since a single set of input variables and parameter values can lead to a range in output values. The variance in model results can also be profoundly affected by stochastic influences. Deterministic models treat rates as fixed items: if the mortality rate is 60%, then exactly 60 out of every 100 individuals will die during that time period, and, by extension, 60.6 die if there are 101 individuals. Therefore, a single set of input parameters gives a single set of values for output variables. In stochastic models, fractions of individuals are not allowed, and such rates are seen as the probability that an individual will die during the time period. The total number that die during any particular period would be a random draw from a binomial distribution (N,p), where N is the total number of individuals and p is the probability. For the example above, while on average the stochastic model will give 60% mortality winter survival, any single realization could give between 0 and 100% mortality with a binomial frequency distribution of survival rates. There are a number of possible events in a model of the evolution of resistance in which such stochastic influences might be incorporated, including mortality (both natural and response to toxins), fecundity, mating and dispersal. These stochastic processes can have a large impact on model variances in some cases and negligible impacts in others (Fig. 4.4a).

In a model simulating bollworm adaptation to *Bt* maize and *Bt* cotton, Storer *et al.* (2003a) conducted five runs of each scenario with a particular parameter set to improve the estimate of the mean values for the end points. Although the standard deviation stabilized after five runs, there was no way to determine whether those five runs were sufficient to capture all of the important possible stochastic variation. For example, stochastic influences can be great when resistance alleles are initially rare, while they are much less important when resistance alleles are relatively common. Selection acts independently of population size while genetic drift becomes more important as population size decreases. Extinction of resistance alleles becomes a real possibility when they are rare, so it is possible that in 99 realizations the resistant allele goes extinct due to genetic drift and resistance effectively does not evolve, while in one realization the resistance allele persists (and resistance may evolve very fast in this case). Clearly the variance in these results is great, and questions addressing the scale of the model become important. If the model system was 100-fold larger (e.g. 100 times the number of fields), would this reduce the variance? Sisterson *et al.* (2004) found that as region size increased the time to resistance decreased, as did the variance in time to resistance. Through the judicious use of spatial

scale parameters, it is possible to transfer variance estimates from between simulations to spatial variance within a single simulation. The appropriateness of
these scale parameters to actual field conditions must be carefully considered
when performing risk assessments.

If a stochastic model is run a sufficient number of times, the variability of the
output can be described by a probability distribution. Just as it is possible to use
repeated realizations of deterministic models with different parameter values to
assess the impact of uncertainty in those values on the variance in model results,
so one can use this technique with a single set of parameters to assess the impact
of stochasticity on the model results. One thousand realizations of a model with
unvarying parameters are likely to provide a reasonable estimate of the influence
of stochastic factors on the evolution of resistance (Fig. 4.4a). While Peck *et al.*
(1999) found that the initial distribution of *Bt* and non-*Bt* fields determined
whether or not resistance foci became established (and therefore resistance
evolved rapidly), Storer (2003) showed in a stochastic spatially explicit model of
corn rootworm adaptation to *Bt* maize that, even if the initial conditions are held
constant, there can be a twofold variation in time to resistance due only to stochasticity in the model.

In this case, we assume that we know all the parameters with no uncertainty
(they remain constant for all runs), and the variance in the results reflects the
limits of our ability to predict the evolution of resistance. It is therefore important
to realize that, short of the impacts of scale factors we noted earlier, this distribution is the limit of our ability to predict the evolution of resistance. No additional knowledge could reduce this variance, just as no additional replication
can reduce the standard deviations in a measurement.

Analysing parameter uncertainty

Resistance modellers have traditionally attempted to understand the importance
of such uncertainty by running a sensitivity analysis: by perturbing parameter
values one at a time and investigating the effect of the perturbation on model
outcomes. Typically, one can ascertain the effect of perturbations on the value
of a parameter by applying ±10% of the default value, while holding all the
other parameters at their default values. By repeating this for each parameter,
the modeller can determine how important the uncertainty around each parameter is to the interpretation of model outcome. For example, Crowder
and Onstad (2005) found that density-dependent mortality and functional dominance of resistance had large effects on rates of adaptation, but fitness costs did
not.

The single parameter perturbation technique only allows the modeller to
determine the effects of changing one parameter at a time and only under
default values for the other parameters. Therefore, the technique does not allow
the modeller to investigate potential interactions among pairs (or even groups)
of parameters. The technique also does not allow the modeller to use information on the magnitude of the uncertainty in parameter values. There may be
more information available about the true value of some parameters than for

others, or certain parameters may be less influenced by environmental effects than others. Storer *et al.* (2003b) attempted to account for differential knowledge among parameters in a sensitivity analysis of their corn earworm/bollworm model by testing ranges of parameter values that represented the authors' opinion of the 'biologically reasonable' range values. The same sensitivity analysis also included investigations into potential interactions among parameters and input variables.

An alternative to the deterministic scenario approach and individual parameter perturbations that has generally been used for risk assessments is to more formally specify our uncertainty in some key model parameters. For each selected parameter, we specify a probability distribution, referred to as the input parameter distribution. In place of running a specific number of scenarios, by subjectively setting the input parameters to some values, in this risk assessment approach one runs a large number of model realizations, randomly drawing new values for each parameter from the specified input probability distributions. After completing many (often greater than 1000) realizations (replicate simulations but with randomly drawn parameters), the results can be arranged as output probability distributions for each model end point, referred to as a Monte Carlo output distribution. They are a direct measure of the impact of parameter uncertainty on the output distributions used for our assessment of risk (Fig. 4.4b, c, and d). One can now begin to address questions such as what is the risk of resistance evolving within a certain amount of time, or what is the probability that a transgenic event will last more than a given number of years. This description of the uncertainty of our results (which differs from our uncertainty in our parameters) is critical to effective risk assessment.

This risk assessment technique effectively assigns probabilities to outcomes and has recently been adopted to examine risk of the evolution of insect resistance to transgenic crops (Maia, 2003; Maia and Dourado-Neto, 2004; Caprio *et al.*, 2006). Uncertainty due to environmental variability can be incorporated into the model by using a different probability each time a parameter is used during a single model run; for example, for each simulated winter and, in a multi-patch model, a different probability for each patch. Uncertainty due to estimates based on field data to provide the true value of a parameter can be accounted for by running a model multiple times, each time using a different value for the parameter (see Fig. 4.3). A large number of parameters can simultaneously be varied in this manner, using random draws for each run for each parameter value, and the model run for hundreds, perhaps thousands, of times. As additional data become available, these uncertainty distributions can be updated.

At this point, the modeller needs to provide probability distributions for the values of the parameters under consideration. Hoffman and Kaplan (1999), discussing characterization of input parameters' uncertainty, pointed out that use of classical statistics (mean, variance) to summarize the variability of direct observations is usually inappropriate for risk projections. They comment that:

> Classical statistics should be restricted to instances when data are obtained from either a random or stratified random design, appropriately

averaged according to the space and time requirements of the assessment, and when the data are directly relevant to the target individuals or populations of interest.

As such situations are not frequent in ecological risk assessments, for which extrapolations are made beyond the spatial extent and time periods in which data have been collected, combining expert knowledge with sampling/experimental data for obtaining input distributions becomes imperative. The form of the probability distribution for any given parameter value should reflect the knowledge of the true value, its variance, and our uncertainty in those values. For example, a parameter about which nothing is known, except the probable range [a,b], could be assigned any value between a and b with equal probability (the uniform [a,b] continuous distribution). Alternatively, if there are maximum (a) and minimum (b) values known, as well as a most likely value, a triangular or beta probability distribution could be used (Fig. 4.3). As many biological input variables are known to fit the normal or log-normal distribution, such distributions can be used if there is some knowledge of the mean and variance. Several alternative probability distributions could also be used if there is sufficient understanding of the parameters to justify their use, including skewed distributions (e.g. gamma, beta, Weibull) or discrete distributions (binomial, Poisson: known probabilities assigned to an enumerable set of possible values). For example, Maia and Dourado-Neto (2004) used uniform, symmetrical triangular, and symmetrical truncated normal distributions for characterizing uncertainty of the resistance allele initial frequency, a key parameter in a simple, deterministic, non-random mating model (Caprio, 1998).

Characterizing input distributions should be regarded as an iterative process: as parameters are identified where uncertainty is critical to model output, additional effort can be used to improve information on those parameters. For more details on the methods for obtaining distributions for uncertain model inputs, see Clemen and Winkler (1999), Hoffman and Kaplan (1999) and Kaplan (2000).

As an example of the Monte Carlo approach to uncertainty analysis, we consider below a very simple deterministic model of resistance evolution. This is a two-patch model, with one patch insecticidal (e.g. a Bt crop) and the other patch non-insecticidal (i.e. refuge), and resistance is conveyed by a single gene with two alleles. The insecticide is assumed to kill 100% of susceptible insects (SS genotype) and none of the resistant insects (RR genotype). Survival of heterozygotes (RS) is given by the functional dominance of the R allele, h. In addition to h, there are two other parameters: the initial frequency of the R allele (q_0) and the proportion of the landscape that is planted to the refuge (x). The population consists of non-overlapping generations, there is no fitness cost associated with resistance, and there is random mating and random oviposition across the two patches (i.e. Hardy–Weinberg dynamics determine the frequency of the three genotypes entering each generation). The model output is the R-allele frequency after ten rounds of selection.

In the first instance, let us assume that nothing is known about the values of h, q_0, or x: that is, we make no assumptions about the frequency of resistance, the functional dominance of resistance, or about how much the landscape will not be treated with the insecticide. Therefore, we assume that the values of each of the parameters are represented by uniform probability distributions between 0.0 and 1.0. The model is run 1000 times, each with a new random independent draw for the values of the three parameters (i.e. Monte Carlo sampling procedure) using Crystal Ball® 2000 (Decisioneering, Inc., Denver, Colorado). The probability distribution of the R-allele frequency after ten rounds of selection is given in Fig. 4.5a. The lowest final R-allele frequency is 0.0051; the highest is 1.0. The mean is 0.85, and the median is 0.95. Seventy-five per cent of the runs result in a final R-allele frequency greater than 0.81; 25% of the runs result in a final R-allele frequency greater than 0.99. We conclude that a large proportion of the three-parameter space leads to rapid resistance evolution. However, we know that probability distribution across the parameter space is not uniform.

In a second iteration of the process, we make some assumptions about the parameter values based on what we know about use of *Bt* crops and resistance to the Cry proteins used in them to better reflect a real-world scenario. *Bt* cotton is planted on 60–90% of cotton grown in the core of the US cotton belt (USDA NASS), varying by geography. Therefore the distribution of values for x, the proportion of the land as refuge (non-*Bt* cotton), can be better characterized by a uniform distribution with a minimum of 10% and a maximum of 40% (we do not have data giving the proportions of *Bt* and non-*Bt* on a finer scale than state level, preventing us from using a more refined probability distribution). The frequency of R alleles to *Bt* cotton in tobacco budworm is believed to be extremely low, since years of monitoring for resistance and use of *Bt* cotton have not found any evidence of resistance. However, we do not believe it is zero. One literature estimate of R-allele frequency is 3×10^{-4} (Gould *et al.*, 1995b). Other modellers typically use a value of 10^{-3} or 10^{-4} as an appropriate estimate. We therefore modify our assumptions about the probability distribution of R-allele frequency to be 10^{-y}, where y is distributed normally with a mean of -3.5 and a standard deviation of 0.5. The functional dominance of resistance to *Bt* cotton in tobacco budworm is also believed to be very low. High levels of resistance to Cry proteins tend to be recessive (Ferré and Van Rie, 2002), and *Bt* cotton expresses a high dose of Cry proteins such that heterozygotes are expected to show very low survival. Functional dominance is unlikely to be zero, and the probability of a given value of h is assumed to be distributed in log-normal manner with a mean value of 0.1 and standard deviation of 0.05.

The Monte Carlo simulation (1000 runs) is conducted using these probability distributions for the three selected parameters to obtain the probability distribution for R-allele frequency after ten rounds of selection (Fig. 4.5b). The distribution is somewhat bimodal, with a large peak at the lowest end of the range and a small peak at the highest end. The lowest final R-allele frequency is 4×10^{-5}; the highest is 0.995. The mean is 0.056, and the median is 0.004. Three-quarters of the runs result in a final R-allele frequency less than 0.02. The refined probabilistic risk assessment based on a very simple deterministic model indicates that the probability of resistance to *Bt* cotton evolving in ten generations in tobacco budworm is extremely low. This exercise has shown us that, while under the worst-case scenario R-allele frequency could theoretically exceed 0.5 within ten generations, under the assumptions of this simplistic

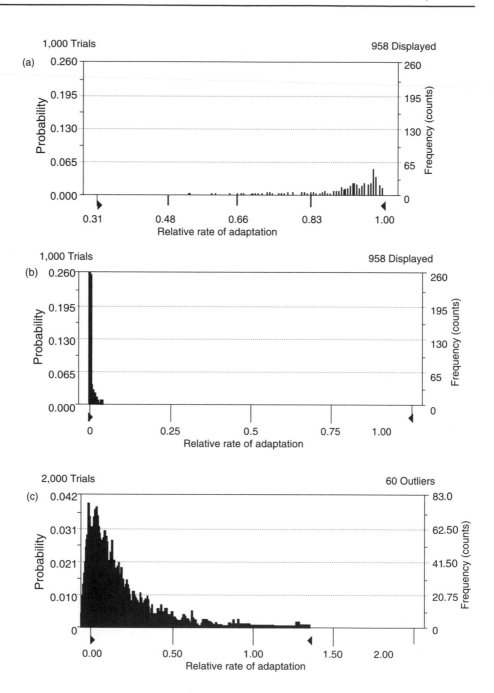

Fig. 4.5. Monte Carlo output distributions for resistance allele frequency after ten rounds of selection (i.e. 10-year gene frequency forecasts from 1000 trials) in a theoretical population based on Monte Carlo parameterization of a simple population genetics model with (a) no prior information and (b) prior, but uncertain, information on selection pressure, dominance of resistance, and initial allele frequencies. Results from 958 trials are displayed in (a), 1000 in (b).

model the probability of this occurring, based on understanding of resistance to *Bt* proteins and *Bt* cotton usage, is <5%.

Similar analyses can be applied to more sophisticated models of resistance evolution that incorporate more biologically realistic processes and more realistic agricultural landscapes. As more parameters are added to a model, it becomes important to consider how they may be correlated with one another. For example, a model may have two parameters governing insect dispersal: the proportion of individuals that disperse at a given time step and the distance that the dispersing insects move during the time step. These parameters may be negatively correlated; the rate of spread of a population across the landscape is likely to be a function of these two parameters. Thus in making random draws for one parameter, one may need to consider the random value drawn for the other, so that the rate of population spread remains realistic. In these situations, we need to use methods for sampling input distributions that account for correlation patterns among parameters (e.g. Latin hypercube sampling) (Iman and Conover, 1982).

These risk assessment techniques also offer advantages in the area of sensitivity analysis. If the randomly assigned parameter values are saved along with the simulation results, a sensitivity analysis that includes interactions and correlations among parameters can be performed. Assuming the data can be linearized, step-wise regression or Bayesian model averaging can be used to identify the parameters or parameter interactions the model is most sensitive to (Caprio *et al*., 2006; McCaffery *et al*., 2006). These interactions are generally left out in conventional scenario-based modelling and sensitivity analysis. For example, in a data set examining the introduction of Vip3A cotton (Caprio and McCaffery, unpublished data), the data were first analysed without interactions between parameters, which showed significant effects for initial gene frequency and dominance of the resistance allele. When all two-way interactions were included into the model, a highly significant interaction term between these two parameters was found. In retrospect, this interaction is reasonable. As we know from basic population genetics texts, the heritability of recessive traits increases dramatically as gene frequency increases, and thus the impact of dominance decreases as initial gene frequency approaches 1.0.

Using a version of a deterministic non-random mating model (Caprio *et al*., 1998), Maia (2003) found complex interactions in a sensitivity analysis. She analysed the sensitivity of two output variables, R-allele frequency across target-pest generations (*RFreq*) and number of generations until resistance (*NGer*), to perturbations in the input parameters initial R-allele frequency and functional dominance of resistance, for different scenarios (combinations of refuge size and refuge pest management). The results showed that *RFreq* sensitivity to both input variables changed considerably among scenarios as well as across generations, ranging from high sensitivity (exponential patterns) in the initial generations to null sensitivity after several rounds of selection.

The model of Western corn rootworm adaptation to *Bt* maize in Storer (2003) was subjected to just such an analysis (Storer, unpublished data). For this Monte Carlo sensitivity analysis, 21 parameters were simultaneously varied according to predefined probability distributions based on assumptions of the underlying distribution of uncertainty or variability for each parameter, and the model was run 2000 times. Two parameter correlations were included. First, survival of susceptible larvae on *Bt* maize was 80% positively correlated with the functional dominance (and therefore with the relative survival of heterozygous larvae). Secondly, the fecundity of adult rootworms was 80% negatively correlated with density-independent mortality in winter, so that the population size in the absence of *Bt* maize remained biologically reasonable. The output from the sensitivity analysis provided a distribution for expected rates of adaptation for rootworms when *Bt* maize was planted on 90% of the acreage (10% refuge). For this model, a relative rate of adaptation (*RRA*) was defined as the average annual increase in R-allele frequency on a log scale, expressed as a fraction of the baseline rate of adaptation:

$$\text{RRA} = \left(\frac{1}{Y}\ln\frac{q_Y}{q_0}\right) / 0.327$$

where q_Y is the R-allele frequency after Y years and 0.327 is the adaptation rate (year^{-1}) for the baseline run (Storer, 2003). This rate is calculated when the R-allele frequency first exceeds 0.075, when the rootworm egg population in the autumn falls below 20,000 per field, or after 10 years, whichever is soonest.

The mean *RRA* for the Monte Carlo analysis was 0.35 (standard deviation = 0.42), while the median was 0.23 (Fig. 4.5c). These averages represent 3.5 to 5.4 times slower adaptation than that predicted using the default parameter settings. The extreme values were −0.86 and 5.07 (negative values of *RRA* resulted from a decline in region-wide r-allele frequency due to local r-allele extinction when populations were small). In the sensitivity analysis, 96.5% of the *RRA* values for the same value were less than 1.24. This finding suggests that the *RRA* value obtained using the default parameters is greater than the 95% confidence limit for the estimate of the true *RRA*.

The parameters with the largest effects on *RRA* were the functional dominance of the R allele (explaining 58.5% of the variation) and the dose (explaining 33.5% of the variation). Lower functional dominance and higher doses caused lower *RRA*. As input parameters, these two parameters had 80% correlation. The next most important parameter was another genetic factor: the fitness cost of resistance (3.1% of the variation). Winter survival (2.9%) and fecundity (1.1%), the pair of correlated parameters, were the next most important. The other parameters combined explained less than 1%. These findings indicate that, until we have isolated resistance alleles in the field, it will be difficult to predict with greater accuracy the rate that the alleles will spread. The probability that adaptation will be slower than predicted by using default values was greater that 95%, suggesting that the set of default parameter values chosen in this case was highly conservative.

Caprio *et al.* (2006) incorporated both stochastic influences and parameter uncertainty in a model of resistance to methyl parathion in Western corn rootworm. They compared the variance in model results (Monte Carlo output distribution for 'time until resistance', *TR*) with parameters fixed at their default values and estimated the impact of stochastic influences. Additional runs of the model integrated both parameter uncertainty in 18 parameters and stochastic influences to estimate their joint influences. In this case, the parameter uncertainties were defined as normal distributions with a standard deviation that was a specified percentage of the default mean value. Uncertainty was not included in Fig. 4.4a. The variability among model output values was entirely due to stochastic demographic processes. For the other cases, the standard deviation increased from 5 to 10 to 20% (Fig. 4.4b, c, and d) demonstrating the impact of increased parameter uncertainty on the *TR* Monte Carlo output distribution. Ultimately, this combination of both stochastic and parametric uncertainty provides an improved description of the model output uncertainty, thus improving our ability to predict the time it will take for resistance to evolve to a particular insecticidal protein in a particular use pattern.

Resistance risk assessment requires that the modeller not only provides point estimates for input parameters, but also a complete characterization of their respective uncertainties via probability distribution functions (PDFs). These PDFs can come from a number of sources, including expert opinion and empirical data (Hoffman and Kaplan, 1999; Kaplan, 2000). In the worst scenario, a flat uniform distribution can be chosen to describe uncertainty of parameters for which there is no prior information. In cases where there is a minimum of data to support a default parameter value, many modellers will be reluctant to speculate on an input probability distribution. Often, however, limits to likely values can be estimated which, together with a most likely value (usually the default value), provide sufficient information for a triangular distribution. The beta distribution can also be used with these three parameters, avoiding the sharp changes in the second moment present in the triangular distribution. The beta distribution also places less emphasis on the tails of the distribution than the triangular distribution. Vose (2000) describes a PERT distribution based on the beta distribution, as well as a modified PERT that reduces the weight placed on the most likely value, gradually transforming into a uniform distribution. The beta distribution requires four parameters: a minimum and maximum value as well as two shape parameters. The modified PERT distribution uses the most likely value and a bias value (defaulting to four in the standard PERT distribution) to determine the two shape parameters. This modified PERT distribution can be used to reflect the modeller's uncertainty in his/her estimate of the most likely value. These distributions require a minimum number of parameters (maximum, minimum, and most likely values) and provide a simple mechanism to incorporate uncertainty into models. Subsequently, sensitivity analysis can indicate which parameters have the most impact on the model results, and further refinements can be made in the uncertainty characterizations for those parameters.

It should be realized that risk estimates can be highly sensitive to the type of input distribution. Using the non-random mating model, Maia and Dourado-Neto (2004) evaluated the influence of three types of distributions for the R-allele initial frequency: uniform (UN), symmetrical triangular (ST), and symmetrical truncated normal (TN), with the same range on risk estimates (probability of *RFreq* exceeding a critical value) across generations for three resistance management scenarios. For all scenarios, they found that risk estimates corresponding to ST and TN were very similar, but estimates obtained considering the UN distribution were quite different, especially for the risks at the initial and last rounds of selection.

Although the risk assessment techniques described attempt to incorporate all available knowledge in a formalized framework, the uncertainty distributions will often contain elements of subjectivity. While the results are often not sensitive to small variations in the maximum or minimum values in a triangular distribution (these are areas of low probability), humans tend to underestimate risk (Vose, 2000; Bedford and Cooke, 2001; Evans and Olsen, 2002), and hence set these extreme values too conservatively. Of particular interest is the process of anchoring (Vose, 2000), in which the modeller anchors his/her estimate to a most likely value and sets extreme values based on this estimate. This usually underestimates uncertainty, and extreme values should be independently estimated. In general, the process of collecting and combining expert opinion requires great attention from the modeller.

In some cases, because of the large number of parameter combinations, it may be difficult to summarize all results using a single criterion, and all combinations may not be reasonable. For example, in Fig. 4.4a, b, c and d (Caprio *et al.*, 2006), one can see a bimodal distribution developing as the uncertainty and observed frequencies of field failures within 3 years increase. In this case, as the variance in the input distributions increased, an increasing number of simulations included parameter combinations that effectively did not control the simulated pest, and the model results reflected only the time required for the initial population to build up to damaging levels. In a subsequent analysis, these values were excluded, because it was presumed that the pesticide never would have been deployed under those settings. Another example is the determination of proper stop criteria that are suitable for all realizations of a model. Most modellers use some sort of frequency-based criterion, for example when the resistance allele frequency in relevant fields exceeds 50%. However, when performing a risk assessment, some of the simulations may include parameters that lead to extremely low rates of selection, and the simulations are generally stopped after a reasonably long period of time (20–50 years). This leads to right censored data with little resolution in how changes in parameters affect resistance evolution beyond the maximum simulated time. Under some situations, this could lead to an underestimate of the sensitivity of the model to certain parameters. Approaches using performance criteria (the ability of the toxin to maintain populations below some level) (Caprio *et al.*, 2006) are also subject to similar problems. An alternative approach is to estimate the exponential rate at which the resistance allele frequency changes per year and to use this as a qualitative measure of risk (e.g. Storer, 2003). The assumption here is that the rate of change in allele frequency is constant as long as resistance alleles are rare (Fig. 4.6). This rate is determined almost entirely by the fitness of resistant heterozygotes relative to that of susceptible homozygotes, and the rate only begins to change when resistant homozygotes begin to be relatively common. As an aside, note that one can change the fitness of the homozygote considerably, changing dominance from recessive to dominant, while having little impact on the rate of resistance evolution. Dominance is relevant only to the degree it describes the

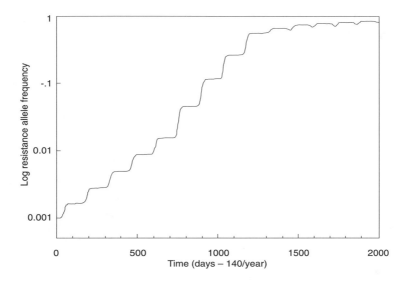

Fig. 4.6. The constant rate of increase in log resistance allele frequency in a simulation of Western corn rootworm resistance to a low-dose transgenic event.

relative fitness of heterozygotes or susceptible homozygotes. The estimated rates can have problems when initial gene frequencies are varied due to uncertainty. In a model of the introduction of Vip3a cotton, Caprio and McCaffery (unpublished data) varied the initial gene frequency from 10^{-1} to 10^{-4}, based on published estimates for other species. In realizations of the model where the initial frequency was close to 10^{-1}, the resistance allele frequency was high enough that it no longer fell within that region where the exponential rate of increase was constant. In these cases the rate could not be reliably estimated, leading to left censored data. Other realizations of the model used parameter values that would lead to long resistance times (indeed, with fitness costs, negative rates can easily be achieved), so frequency-based criteria would also lead to loss of information due to right censored data. It may not be unusual that the extensive variety of results in a risk assessment will require that some information be lost, but the results should indicate the important parameters within the time frame examined.

Resistance simulations, especially prospective ones run before widespread deployment of a new technology, tend to be run assuming maximum adoption of the technology. While it is possible to predict technology adoption curves, this is rarely done, meaning that the results are highly conservative. One outcome of this approach may be to magnify the effects of parameter uncertainty on the predictions of resistance evolution.

Conclusions: Towards the Appropriate Interpretation and Use of Resistance Models

As resistance risk assessment and management has developed since the 1950s, especially spurred on by the release of transgenic insecticidal crops, we have seen enormous reliance on models. Empirical experimentation with insect resistance is inherently tricky, since a successful field experiment to examine the effectiveness of a given strategy would inevitably create the problem we are trying to avoid. Experiments in greenhouses (Tang *et al.*, 2001; Zhao *et al.*, 2003) can be regarded as models of the field in that the environmental influences are controlled, as are many of the biological processes along with the operational manipulations. Data from such experiments have been extremely useful in validating the findings of the simple generalized computer models, which can be regarded as experiments in cyberspace. From these physical and virtual experiments, we have been able to assess the relative merits of different resistance management strategies, such as high-dose/refuge and gene pyramiding. However, it remains problematic to extend these findings to devise appropriate strategies to use in commercial field conditions.

More complex models are more system-specific and have greater predictive utility to understand the resistance risks in any particular use pattern. However, as we have seen, it is vital to understand the predictive limitations of these models, particularly by understanding the uncertainty associated with the model predictions. 'Garbage in, garbage out' is a truism of these models, and modellers are faced with a huge challenge to parameterize them correctly. Indeed, there is generally no single correct parameter set to use when modelling across space and time.

Crude attempts to capture the uncertainty by running best-case and worst-case scenarios are of limited utility if there is no attempt to relate those scenarios to the real world. In this chapter, we have described the application of established risk assessment tools to resistance management models that provide more useful descriptions of our uncertainties in predictions of resistance evolution. While simple models have been used to develop regulatory policy around resistance management (ILSI/HESI, 1999; US EPA, 2001; Fitt *et al.*, 2006), we are seeing with more sophisticated approaches that not all the important processes are necessarily taken into account, and that the resistance risk is not uniform for all products and all use patterns. That resistance has not evolved to *Bt* crops after more than 10 years of use is evidence that one or more of the assumptions made in the early models were highly conservative or that those models inadequately described the resistance risks (Tabashnik *et al.*, 2003).

System-specific models with an appropriate uncertainty analysis in a risk assessment context show whether or not resistance management strategies are needed, and, if so, what strategies should be implemented. Clearly, the need for flexibility inherent to this approach does not lend itself to a 'one size fits all' regulatory-driven strategy, as much as it does to a locally developed and implemented strategy within an overall regulatory policy. With probabilistic risk assessment tools, we can start to determine under what circumstances there may be an unreasonable resistance risk, and consequently under what circumstances resistance management is warranted. Of course, for implementation of the risk assessment process to be useful, there must be some agreement of what constitutes an unacceptable risk. By leaving this undefined, the European and Mediterranean Plant Protection Organization has left the interpretation of unacceptability to the subjective opinion of regulatory decision makers (Leonard, 2000). For resistance to *Bt* crops, this could be failure of the product within a certain time frame with a certain probability. With spatial models, this could be further refined by stating over what proportion of the area where the *Bt* crop is used resistance would be unacceptable. A regulatory policy developed around defining these unacceptable risks would go a long way towards establishing rational resistance management plans.

References

Alstad, D.N. and Andow, D.A. (1995) Managing the evolution of insect resistance to transgenic plants. *Science* 268, 1894–1896.

Barber, G.W. (1936) *The Cannibalistic Habits of the Corn Ear Worm*. Technical Bulletin No. 499. US Department of Agriculture, Washington, DC.

Bedford, T and Cooke, R. (2001) *Probabilistic Risk Analysis: Foundations and Methods*. Cambridge University Press, Cambridge, UK.

Caprio, M.A. (1994) *Bacillus thuringiensis* gene deployment and resistance management in single and multi-tactic environments. *Biocontrol Science and Technology* 4, 487–497.

Caprio, M.A. (1998) The evolution of resistance: a simple deterministic model. http://www.msstate.edu/Entomology/PGjava/ILSImodel.html (accessed December 2007).

Caprio, M.A. (2001) Source–sink dynamics between transgenic and non-transgenic habitats and their role in the evolution of resistance. *Journal of Economic Entomology* 94, 698–705.

Caprio, M.A. and Tabashnik, B.E. (1992) Gene flow accelerates local adaptation among finite populations: simulating the evolution of insecticide resistance. *Journal of Economic Entomology* 85, 611–620.

Caprio, M.A., Faver, M.K. and Hankins, G. (2004) Evaluating the impacts of refuge width on source–sink dynamics between transgenic and non-transgenic habitats. *Journal of Insect Science* 4, 3.

Caprio, M.A., Nowatzki, T., Siegfried, B., Meinke, L.J., Wright, R.J. and Chandler, L.D. (2006) Assessing risk of resistance to aerial applications of methyl-parathion in western corn rootworm (Coleoptera: Chrysomelidae). *Journal of Economic Entomology* 99, 483–493.

Carrière, Y.P., Dutilleul, P., Ellers-Kirk, C., Pedersen, B., Haller, S., Antilla, L., Dennehy, T.J. and Tabashnik, B.E. (2002) Sources, sinks, and the zone of influence of refuges for managing insect resistance to *Bt* crops. *Ecological Applications* 14, 1615–1623.

Chapman, R.F. (1982) *The Insects: Structure and Function*. Harvard University Press, Cambridge, Massachusetts.

Clemen, R.T. and Winkler, R.L. (1999) Combining probability distributions from experts in risk analysis. *Risk Analysis* 19, 187–203.

Comins, H.N. (1977a) The development of insecticide resistance in the presence of migration. *Journal of Theoretical Biology* 64, 177–197.

Comins, H.N. (1977b) The management of pesticide resistance. *Journal of Theoretical Biology* 65, 399–420.

Comins, H.N. (1979) Analytic methods for the management of pesticide resistance. *Journal of Theoretical Biology* 77, 171–188.

Crow, J.F. (1957) Genetics of insect resistance to chemicals. *Annual Review of Entomology* 2, 227–246.

Crowder, D.W. and Onstad, D.W. (2005) Using a generational time-step model to simulate dynamics of adaptation to transgenic corn and crop rotation by western corn rootworm (Coleoptera: Chrysomelidae). *Journal of Economic Entomology* 98, 518–533.

Curtis, C.F., Cook, L.M. and Wood, R.J. (1978) Selection for and against insecticide resistance and possible methods of inhibiting the evolution of resistance in mosquitoes. *Ecological Entomology* 3, 273–287.

Davis, P.M. and Onstad, D.W. (2000) Seed mixtures as a resistance management strategy for European corn borers (Lepidoptera: Crambidae) infesting transgenic corn expressing Cry1Ab protein. *Journal of Economic Entomology* 93, 937–948.

Evans, J.R. and Olsen, D.L. (2002) *Introduction to Simulation and Risk Analysis*, 2nd edn. Prentice-Hall, Upper Saddle River, New Jersey.

Ferré, F. and Van Rie, J. (2002) Biochemistry and genetics of insect resistance to *Bacillus thuringiensis*. *Annual Review of Entomology* 47, 501–533.

Fitt, G.P., Omoto, C., Maia, A.H.N., Waquil, J.M., Caprio, M., Okech, M.A., Cia, E., Huan, N.H. and Andow, D. (2006) Resistance risks of *Bt* cotton and their management in Brazil: environmental risk assessment of genetically modified organisms. In: Hilbeck, A., Andow, D. and Fontes, E.M.G. (eds) *Methodologies for Assessing Bt Cotton in Brazil*, Vol. 2. CAB International, Wallingford, UK, pp. 300–345.

Georghiou, G.P. and Taylor, C.E. (1977a) Genetic and biological influences in the evolution of insecticide resistance. *Journal of Economic Entomology* 70, 319–323.

Georghiou, G.P. and Taylor, C.E. (1977b) Operational influences in the evolution of insecticide resistance. *Journal of Economic Entomology* 70, 653–658.

Gould, F. (1986) Simulation models for predicting durability of insect-resistant germplasm: hessian fly (Diptera: Cecidomyiidae)-resistant winter wheat. *Environmental Entomology* 15, 11–23.

Gould, F. (1994) Potential problems with high-dose strategies for pesticidal engineered crops. *Biocontrol Science and Technology* 4, 451–462.

Gould, F. (1998) Sustainability of transgenic insecticidal cultivars: integrating pest genetics and ecology. *Annual Review of Entomology* 43, 701–726

Gould, F., Kennedy, G.G. and Johnson, M.T. (1991) Effects of natural enemies on the rate of herbivore adaptation to resistant host plants. *Entomologia Experimentalis et Applicata* 58, 1–14.

Gould, F., Anderson, A., Reynolds, A., Bumgarner, L. and Moar, W. (1995a) Selection and genetic analysis of a *Heliothis virescens* (Lepidoptera: Noctuidae) strain with high levels of resistance to *Bacillus thuringiensis* toxins. *Journal of Economic Entomology* 88, 1545–1559.

Gould, F., Anderson, A., Jones, A., Sumerford, D., Heckel, D.E., Lopez, J., Micinski, S., Leonard, R. and Laster, M. (1995b) Initial frequency of alleles for resistance to *Bacillus thuringiensis* toxins in field populations of *Heliothis virescens*. *Proceedings National Academy of Science USA* 94, 3519–3523.

Guse, C.A., Onstad, W., Buschman, G.L.L., Porter, G.P., Higgins, R.A., Sloderbeck, P.E., Cronholm, G.B. and Peairs, F.B. (2002) Modeling the development of resistance by stalk-boring *Lepidoptera* (Crambidae) in areas with irrigated transgenic corn. *Environmental Entomology* 31, 676–685.

Gustafson, D.I., Head, G.P. and Caprio, M.A. (2006) Modeling the impact of alternative hosts on *Helicoverpa zea* adaptation to bollgard cotton. *Journal of Economic Entomology* 99, 2116–2124.

Hoffman, F.O. and Kaplan, S. (1999) Beyond the domain of direct observation: how to specify a probability distribution that represents the 'state of knowledge' about uncertainty inputs. *Risk Analysis* 19, 131–134.

ILSI/HESI (1999) *An Evaluation of Insect Resistance Management in Bt Field Corn: A Science-Based Framework for Risk Assessment and Risk Management. Report of an Expert Panel.* International Life Sciences Institute/Health and Environmental Sciences Institute, Washington, DC.

Iman, R.L. and Conover, W.J. (1982) A distribution free approach to inducing rank correlations among input variables. *Communications in Statistics: Simulation and Computation* 11, 311–334.

Isukapalli, S.S. and Georgopoulus, P.G. (2001) Computational Methods for Sensitivity and Uncertainty Analysis for Environmental and Biological Models. http://www.ccl.rutgers.edu/reports/EPA/edmas_v3_epa.pdf (accessed November 2007).

Jurat-Fuentes, J.L., Gould, F.L. and Adang, M.J. (2000) High levels of resistance and cross-resistance to *Bacillus thuringiensis* Cry1 toxins in *Heliothis virescens* are due to reduced toxin binding and pore formation. *Resistant Pest Management* 11, 23–24.

Kaplan, S. (2000) Combining probability distributions from experts in risk analysis. *Risk Analysis* 20, 155–156.

Kennedy, G.G. and Storer, N.P. (2000) Life systems of polyphagous arthropod pests in temporally unstable cropping systems. *Annual Review of Entomology* 45, 467–493.

Lenormand, T. and Raymond, M. (1998) Resistance management: the stable zone strategy. *Proceedings of the Royal Society of London. Series B* 265, 1985–1990.

Leonard, P.K. (2000) Resistance risk evaluation, 'a European regulatory perspective'. *Crop Protection* 19, 905–909.

Mallet, J. and Porter, P. (1992) Preventing insect adaptation to insect-resistant crops: are seed mixtures or refugia the best strategy? *Proceedings: Biological Sciences* 250, 165–169.

Maia, A.H.N. (2003) Modelagem da evolução da resistência de pragas a toxinas Bt expressas em culturas transgênicas: quantificação de risco utilizando análise de incertezas. PhD thesis, Universidade de São Paulo, Brazil; http://www.teses.usp.br/teses/disponiveis/11/11136/tde-19012004-100211

Maia, A.H.N. and Dourado Neto, D. (2004) Probabilistic tools for assessment of pest resistance risk associated to insecticidal transgenic crops. *Scientia Agricola* 61, 481–485.

McCaffery, A., Caprio, M., Jackson, R., Marcus, M., Martin, T., Dickerson, D., Negrotto, D., O'Reilly, D., Chen, E. and Lee, M. (2006) Effective IRM with the novel insecticidal protein Vip3A. In: *Proceedings of the 2006 Beltwide Cotton Production Research Conference.* National Cotton Council, San Antonio. Texas.

McGaughey, W.H. and Whalon, M.E. (1992) Managing insect resistance to *Bacillus thuringiensis* toxins. *Science* 258, 1451–1455.

Metcalf, R.L. (1973) A century of DDT. *Journal of Agriculture and Food Chemistry* 21, 511–519.

Onstad, D.W. and Gould, F. (1998) Modeling the dynamics of adaptation to transgenic maize by European corn borer (Lepidoptera: Pyralidae). *Journal of Economic Entomology* 91, 585–593.

Onstad, D.W., Guse, C.A., Spencer, J.L., Levine, E. and Gray, M.E. (2001) Modeling the dynamics of adaptation to transgenic corn by western corn rootworm (Coleoptera: Chrysomelidae). *Journal of Economic Entomology* 94, 529–540.

Parker, C.D. Jr (2000) Temporal distribution of Heliothines in corn-cotton cropping systems of the Mississippi Delta and relationship to yield and population growth. PhD Dissertation, Mississippi State University, Mississippi State, Mississippi.

Peck, S.L. (2004) Simulation as experiment: a philosophical reassessment for biological modeling. *Trends in Ecology & Evolution* 19, 530–534.

Peck, S.L., Gould, F. and Ellner, S.P. (1999) Spread of resistance in spatially extended regions of transgenic cotton: implications for management of *Heliothis virescens* (Lepidoptera: Noctuidae). *Journal of Economic Entomology* 92, 1–16.

Plapp, F.W. Jr, Browning, C.R. and Sharpe, P.J.H. (1979) Analysis of the rate of development of insecticide resistance based on simulation of a genetic model. *Environmental Entomology* 8, 494–500.

Richmond, B. (2001) *An Introduction to Systems Thinking: Stella Software*. High Performance Systems, Hanover, New Hampshire.

Roush, R.T. (1994) Managing pests that their resistance to *Bacillus thuringiensis*: can transgenic crops be better than sprays? *Biocontrol Science and Technology* 4, 501–516.

Siqueira, H.A., Moellenbeck, D., Spencer, T. and Siegfried, B.D. (2004) Cross-resistance of Cry1Ab-selected *Ostrinia nubilalis* (Lepidoptera: Crambidae) to *Bacillus thuringiensis* -endotoxins. *Journal of Economic Entomology* 97, 1049–1057.

Sisterson, M.S., Antilla, L., Carrière, Y., Ellers-Kirk, C. and Tabashnik, B.E. (2004) Effects of insect population size on evolution of resistance to transgenic crops. *Journal of Economic Entomology* 97, 1413–1424.

Sisterson, M.S., Carrière, Y., Dennehy, T.J. and Tabashnik, B.E. (2005) Evolution of resistance to transgenic crops: interactions between insect movement and field distribution. *Journal of Economic Entomology* 98, 1751–1762.

Stadelbacher, E.A. (1979) *Geranium dissectum*: an unreported host of the tobacco budworm and bollworm and its role in their seasonal and long term population dynamics in the delta of Mississippi. *Environmental Entomology* 8, 1153–1156.

Storer, N.P. (2003) A spatially explicit model simulating western corn rootworm (Coleoptera: Chrysomelidae) adaptation to insect-resistant maize. *Journal of Economic Entomology* 96, 1530–1547.

Storer, N.P., Peck, S.L., Gould, F., Van Duyn, J.W. and Kennedy, G.G. (2003a) Spatial processes in the evolution of resistance in *Helicoverpa zea* (Lepidoptera: Noctuidae) to *Bt* transgenic corn and cotton in a mixed agroecosystem: a biology-rich stochastic simulation model. *Journal of Economic Entomology* 96, 156–172.

Storer, N.P., Peck, S.L., Gould, F., Van Duyn, J.W. and Kennedy, G.G. (2003b) Sensitivity analysis of a spatially-explicit stochastic simulation model of the evolution of resistance in *Helicoverpa zea* (Lepidoptera: Noctuidae) to *Bt* transgenic corn and cotton. *Journal of Economic Entomology* 96, 173–187.

Tabashnik, B.E. (1986) Computer simulation as a tool for pesticide resistance management. In: *Pesticide Resistance: Strategies and Tactics for Management*. National Academy Press, Washington, DC, pp. 194–206.

Tabashnik, B.E. (1994a) Delaying insect adaptation to transgenic plants: seed mixtures and refugia reconsidered. *Proceedings of the Royal Society of London. Series B* 255, 7–12.

Tabashnik, B.E. (1994b) Evolution of resistance to *Bacillus thuringiensis*. *Annual Review of Entomology* 39, 47–79.

Tabashnik, B.E. and Croft, B.A. (1982) Managing pesticide resistance in crop–arthropod complexes: interactions between biological and operational factors. *Environmental Entomology* 11, 1137–1144.

Tabashnik, B.E. and Croft, B.A. (1985) Evolution of pesticide resistance in apple pests and their natural enemies. *BioControl* 30, 37–49.

Tabashnik, B.E., Carrière, Y., Dennehy, T.J., Morin, S., Sisterson, M.S., Roush, R.T., Shelton, A.M. and Zhao, J.Z. (2003) Insect resistance to transgenic *Bt* crops: lessons from the laboratory and field. *Journal of Economic Entomology* 96, 1031–1038.

Tang, J.D., Collins, H.L., Metz, T.D., Earle, E.D., Zhao, J.Z., Roush, R.T. and Shelton, A.M. (2001) Greenhouse tests on resistance management of *Bt* transgenic plants using refuge strategies. *Journal of Economic Entomology* 94, 240–247.

Taylor, C.E. (1983) Evolution of resistance to insecticides: the role of mathematical models and computer simulation. In: Georghiou, G.P. and Saito, T. (eds) *Pest Resistance to Pesticides*. Plenum Press, New York, pp. 163–173.

Taylor, C.E. (1986) Genetics and the evolution of resistance to insecticides. *Biological Journal of the Linnean Society* 27, 103–112.

Taylor, C.E. and Georghiou, G.P. (1979) Suppression of insecticide resistance by alteration of gene dominance and migration. *Journal of Economic Entomology* 72, 105–109.

Taylor, C.E. and Georghiou, G.P. (1982) Influence of pesticide persistence in evolution of pesticide resistance. *Environmental Entomology* 11, 746–750.

Taylor, C.E. and Georghiou, G.P. (1983) Evolution of resistance insecticides: a case study on the influence of migration and insecticide decay rate. *Journal of Economic Entomology* 76, 704–707.

US Department of Agriculture, National Agriculture Statistics Service USDA NASS. Quick Stats: Agricultural Statistics Database. http://www.nass.usda.gov/QuickStats

US EPA (2001) Biopesticides Registration Action Document – *Bacillus thuringiensis* Plant-Incorporated Protectants. D. Insect resistance management. http://www.epa.gov/pesticides/biopesticides/pips/bt_brad.htm (accessed November 2007).

Vose, D. (2000) *Risk Analysis: A Quantitative Guide*. Wiley & Sons, New York.

Wearing, C.H. and Hokkanen, H.M T. (1995) Pest resistance to *Bacillus thuringiensis*: ecological crop assessment for *Bt* gene incorporation and strategies of management. In: Hokkanen, H.M.T. and Lynch, J.M. (eds) *Biological Control: Benefits and Risks*. Cambridge University Press, Cambridge, UK, pp. 236–252.

Zhao, J.Z., Cao, J., Li, Y., Collins, H.L., Roush, R.T., Earle, E.D. and Shelton, A.M. (2003) Transgenic plants expressing two *Bacillus thuringiensis* toxins delay insect resistance evolution. *Nature Biotechnology* 21, 1493–1497.

5 Pesticide and Transgenic Plant Resistance Management in the Field

D.A. Andow[1], G.P. Fitt[2], E.J. Grafius[3], R.E. Jackson[4], E.B. Radcliffe[1], D.W. Ragsdale[1] and L. Rossiter[5]

[1]Department of Entomology, University of Minnesota, St. Paul, Minnesota, USA; [2]CSIRO Long Pocket Laboratories, Indooroopilly, Queensland, Australia; [3]Department of Entomology, Michigan State University, East Lansing, Michigan, USA; [4]USDA–ARS, Stoneville, Mississippi, USA; [5]NSW Department of Primary Industries, Australian Cotton Research Institute, Narrabi, New South Wales, Australia

5.1 Introduction

As summarized in other chapters of this book, resistance evolution in insects to pesticides has been recognized as an agricultural problem since the early 1900s. However, it took most of the 20th century before an entomological consensus was reached about the seriousness of the problem (NRC, 1986). Whalon et al. (2007) (Chapter 1, this volume) report 7637 cases of resistance to particular pesticidal products. Using these data, we found that 16 species of arthropods account for 3237 (43%) of these cases (Table 5.1). These include three acari, a cockroach, two aphids, a whitefly, two beetles, three lepidoptera, three mosquitoes and the housefly.

Given the seriousness of the problem, what measures have been taken to reduce its severity? What practices are used in the field to manage the risk of resistance evolution? What measures are taken to delay the onset of resistance failures? The goal of insect resistance management (IRM) is to delay or prevent the field occurrence of resistance failures by delaying or preventing the evolution of resistance. IRM can be characterized as either responsive or pre-emptive. Responsive strategies react to the widespread occurrence of field resistance, while pre-emptive strategies attempt to avoid or delay resistance before it occurs in the field (Brown, 1981; Dennehy, 1987; Sawicki and Denholm, 1987).

In this chapter, we use a case study approach to identify potential generalizations about IRM, barriers to field implementation and research needs for the future. We focus on three of the agricultural pests with the most cases of resistance to pesticides: (i) Colorado potato beetle, *Leptinotarsa decemlineata* (Say) (Coleoptera: Chrysomelidae); (ii) green peach aphid, *Myzus persicae* (Sulzer)

Table 5.1. Species with the highest reported number of cases of resistance. (From Whalon *et al.*, 2005; http://www.pesticideresistance.org/DB/index.html)

Species	Family – Order	Common name	Cases
Panonychus ulmi	Tetranychidae – Acari	European red mite	178
Tetranychus urticae	Tetranychidae – Acari	Two-spotted spider mite	327
Boophilus microplus	Ixodidae – Acari	Southern cattle tick	127
Blattella germanica	Blattellidae – Orthoptera	German cockroach	213
Aphis gossypii	Aphididae – Homoptera	Melon and cotton aphid	103
Myzus persicae	Aphididae – Homoptera	Green peach aphid	293
Bemisia tabaci	Aleyrodidae – Homoptera	Sweet potato whitefly	167
Leptinotarsa decemlineata	Chrysomelidae – Coleoptera	Colorado potato beetle	175
Tribolium castaneum	Tenebrionidae – Coleoptera	Red flour beetle	108
Plutella xylostella	Plutellidae – Lepidoptera	Diamondback moth	278
Helicoverpa armigera	Noctuidae – Lepidoptera	Cotton bollworm	435
Heliothis virescens	Noctuidae – Lepidoptera	Tobacco budworm	106
Aedes aegypti	Culicidae – Diptera	Yellow fever mosquito	196
Culex pipiens pipiens	Culicidae – Diptera	House mosquito	119
Culex quinquefasciatus	Culicidae – Diptera	Southern house mosquito	229
Musca domestica	Muscidae – Diptera	Housefly	183

(Homoptera: Aphididae); and (iii) cotton bollworm, *Helicoverpa armigera* (Hübner) (Lepidoptera: Noctuidae). IRM is strongly encouraged in potato, because resistance is so common in Colorado potato beetle and green peach aphid, not to mention pea leaf miner, *Liriomyza huidobrensis* (Blanchard), cotton aphid, and palm thrips, *Thrips palmi* (Karny). Intriguingly, except for tobacco budworm (*Heliothis virescens* (F.)) (*Lepidoptera: Noctuidae*), diamondback moth (*Plutella xylostella*) and housefly (*Musca domestica*), we were unable to find much published literature for the other species in Table 5.1 which addressed resistance management. In addition to these three species, we also summarize progress on IRM for the transgenic *Bt* crops, *Bt* cotton and *Bt* maize. Transgenic crops have attracted considerable attention, and nearly all of the insecticidal transgenic crops have been commercialized with an IRM plan. In many cases, an IRM plan was a major requirement for obtaining commercial registration of these transgenic crops. A comparison of the state of practice of resistance management for pesticides and transgenic crops will be illuminating.

5.2 Insect Resistance Management for Insecticides

Colorado potato beetle, *Leptinotarsa decemlineata*

Colorado potato beetle has developed resistance to all synthetic insecticides used for its control, including organophosphates, organochlorines, carbamates and pyrethroids (Bishop and Grafius, 1996; Whalon *et al.*, 2005). During a period of severe resistance problems (1990–1994), the Michigan potato industry experienced crop losses and control costs of $350–615/ha per annum (Grafius,

1997). Growers in regions with severe resistance problems adopted some creative management practices including propane flamers for control of early-season adults (Moyer *et al.*, 1992) and plastic-lined trenches to reduce migration between fields (Misener *et al.*, 1993).

The neonicotinoid imidacloprid was approved for use in potatoes in 1995 and was rapidly adopted throughout the eastern and north central potato production regions of North America, resulting in a net decrease of active ingredient of insecticides used on potatoes (MIASS, 1994, 1995/1996; Guenthner *et al.*, 1999) and cessation of mechanical controls. Currently, 70–80% of the potato acreage in the eastern and north central USA is treated with imidacloprid or thiamethoxam for Colorado potato beetle control (NASS, 2004). Most applications are at planting in furrow applications or seed treatments. There has been a rapid increase in the number of neonicotinoid products, and there were seven active ingredients used in a dozen or more registered products available to US potato producers in 2007.

Resistance to imidacloprid in Colorado potato beetle in the field was first found in 1997 (Zhao *et al.*, 2000). It is now widespread throughout the eastern production region (Bishop *et al.*, 2003; Grafius *et al.*, 2004) and appeared for the first time in the north central production region in 2004 (Grafius *et al.*, 2005). Resistance to thiamethoxam, another neonicotinoid, was first found in Colorado potato beetle in the field in 2003 (Byrne *et al.*, 2004).

The Neonicotinoid Subcommittee of the US Environmental Protection Agency's (EPA) Insecticide Resistance Action Committee (IRAC) has drafted overall neonicotinoid resistance management recommendations (IRAC, 2005). The US National Potato Council has also developed neonicotinoid resistance management recommendations for Colorado potato beetle and green peach aphid in potatoes (NPC, 2005). Both groups recommend use of neonicotinoids within a broad integrated pest management (IPM) programme and recommend avoiding repeated use of neonicotinoid, insecticides, such as a foliar neonicotinoid, following an application of a neonicotinoid at planting. The underlying principles are in consideration of the entire pest complex, reducing the need to apply insecticides and rotating insecticide modes of action, so that an insecticide with the same mode of action is not used sequentially on the pest. Being a specialized pest of potato and other Solanaceae, there are generally no large refuge habitats for Colorado potato beetle outside potato fields.

Pest management options for Colorado potato beetle are limited (Hare, 1990). There are no biological control agents that offer effective control of Colorado potato beetle. Likewise, no commercial potato varieties offer host plant resistance; NewLeaf™, a genetically engineered resistant potato variety, was removed from the market in 2000. Scouting and the use of economic thresholds are not applicable for managing Colorado potato beetle when neonicotinoids are used, because they are applied typically at planting. However, scouting and thresholds are needed for pests not controlled by neonicotinoids and to detect Colorado potato beetle control failures and potential resistance problems.

The primary IRM tactic to reduce the need to apply insecticides for Colorado potato beetle is crop rotation. In addition, crop rotation forces dispersal of beetles and may increase genetic mixing. Although crop rotation may accelerate resistance evolution in highly mobile pests (Peck *et al.*, 1999), the biology of Colorado potato beetle suggests that rotation would be beneficial for IRM in this case. Crop rotation delays arrival of beetles into a new potato field compared with a non-rotated field and reduces beetle population size (Wright, 1984; Weisz *et al.*, 1994, 1996), reducing the need for insecticide treatment. Without rotation, there may be little between-field movement of beetles. Imidacloprid resistance in Colorado potato beetle is thought to be inherited in an incompletely recessive manner (Zhao *et al.*, 2000), and the dispersal and gene flow induced by crop rotation may be important to delay resistance evolution.

Although 75% of Michigan potato growers use crop rotation (Michigan Potato Industry Commission, 1998), in most cases this has had minimal impact on Colorado potato beetle populations. To significantly reduce beetle populations, rotated fields should be >500 m apart causing beetles to fly instead of walk between fields (Weisz *et al.*, 1996). However, the primary objective of crop rotation is usually disease management, and, often, fields are only rotated side-by-side. Crop rotation is also limited by available land and irrigation. Some growers alternate sides of a centre pivot in a two-year rotation. Planning between neighbouring growers also needs to be considered (Weisz *et al.*, 1996). In a 1994 survey of Ohio growers, fewer than half the potato fields were rotated >400 m (Waller *et al.*, 1998). Thus, rotation has not been an effective, widespread IRM practice.

Another pest management practice that is commonly combined with side-by-side crop rotation is selective treatment of field borders. Borders adjacent to the previous year's potato crop are often the most heavily infested portion of a field (Dively *et al.*, 1998; Hoy *et al.*, 2000), because most beetles walk into the new field and remain on border rows. Border sprays can help IRM by leaving a portion of the population (beetles in the middle of the field) untreated, compared with whole-field treatment. Impact of border sprays on IRM depends on the proportion of the population exposed to insecticide application. IRM practices, such as trap crops and mechanical controls (propane flamers, plastic-lined trenches and crop vacuums), were used on <3 and 0% of crop acreage, respectively, after adoption of neonicotinoids (Michigan Potato Industry Commission, 1998).

If a second insecticide application is necessary after that applied at planting, the IRM practice is to use an insecticide with a different mode of action. Treatment with imidacloprid at planting followed by foliar treatment with imidacloprid was a common practice early on, but now most growers rotate insecticides. The recent availability of spinosad and increasing imidacloprid resistance problems have encouraged growers to use spinosad for later-season control of larvae and subsequent summer adults if treatment is needed following at-planting application of a neonicotinoid.

Green peach aphid, *Myzus persicae*

IRM programmes exist for a limited number of aphid species and cropping systems worldwide. Insecticide resistance has been best studied in the green peach aphid. Multiple mechanisms of resistance have been identified (Devonshire *et al.*, 1998), including gene amplification (*E4* and *FE4*), insensitive or modified acetylcholinesterase (MACE), and knockdown resistance (kdr) that impart resistance to organophosphate, carbamate and pyrethroid insecticides, respectively. Resistance to organophosphates was first documented in green peach aphid populations from Washington State in 1952 (Anthon, 1955). Resistant populations spread rapidly, and, within 10 years, green peach aphid populations with identical *E4* genes were found in Europe and other parts of the world (Devonshire *et al.*, 1998).

Rarely does green peach aphid reach densities that result in direct plant damage, but the species is of great economic importance as a vector of over 100 plant viruses, 13 in potatoes alone (Blackman and Eastop, 2000). As virus vectors, even low aphid densities can result in an unacceptable level of disease incidence. To limit disease spread in crops, growers often apply multiple insecticide applications to achieve season-long aphid control, which has resulted in intense selection for insecticide resistance.

For green peach aphid, there is a limited number of insecticides worldwide that still provide acceptable control. Among these is the organophosphate methamidophos, a compound that has been registered for use in North America since 1972, but is so toxic to birds and mammals that it is used reluctantly by many potato producers. The neonicotinoids are active and persistent at low rates (g/ha) when applied to seed or soil. A novel insecticide that gives acceptable green peach aphid control is pymetrozine (class pyridine azomethine). The mode of action of pymetrozine remains unknown, but apparently it is a central nervous system toxin. Insecticide-resistant green peach aphid is readily controlled with this material (Davis *et al.*, 2003). Once exposed to pymetrozine, green peach aphid immediately withdraws its stylets from the plant and never again attempts to feed (Harrewijn and Kayser, 1997).

Because green peach aphid has such a high propensity to develop insecticide resistance, it is imperative that IRM programmes be implemented to preserve the few effective materials still available. There are isolated reports of resistance to methamidophos (Whalon *et al.*, 2005), but there are no reported cases of resistance to any neonicotinoid or pymetrozine in green peach aphid (Nauen and Elbert, 2003). Because so many neonicotinoid-based insecticides are available for use, it is an educational challenge to advise crop professionals to practise sound IRM.

To slow the development of resistance in green peach aphid, it is essential to consider the entire pest complex associated with a crop so that IRM practices do not interfere with pest management needs. IRM for any insect aims to reduce the exposure of the target pest to insecticides in three general ways: (i) by ensuring widespread refuges, which are habitats where the pest is not exposed to the insecticide; (ii) by minimizing the need for insecticide applications; and (iii) by avoiding sequential use of insecticides with the same mode of action.

For green peach aphid, a species with 875 known hosts (Leonard *et al.*, 1970), extensive refuges undoubtedly exist outside potato. However, because sexual reproduction is limited in green peach aphid, it is unclear whether refuges can slow development of insecticide resistance in this species. With holocyclic green peach aphid, sexual reproduction occurs only once at the end of the growing season, while 15 or more generations of parthenogenetic reproduction occur during the season. In permissive climates, the situation is further complicated by the occurrence of anholocyclic populations overwintering as asexual clones.

Avoiding sequential use of insecticides with the same mode of action may be an effective IRM strategy. Studies in the UK have shown that anholocyclic, organophosphate-resistant clones are less likely to survive the winter than susceptible clones (Devonshire *et al.*, 1998), which suggests a fitness cost for maintaining resistance in the absence of selection pressure. If fitness costs are common in resistant green peach aphid, avoiding sequential use of insecticides with the same mode of action could allow reversion to susceptibility.

When many control options are available, practitioners may require guidance to avoid inadvertent sequential use of products with the same mode of action. The IRAC, an inter-company committee (www.irac-online.org), has developed a classification system of insecticides, placing them into 28 groups based on their mode of action. The classification allows for easier implementation of IRM, because practitioners need not have knowledge of insecticide chemistry to avoid sequential use of products with the same mode of action.

With respect to IRM, North American potato production presents major regional differences (Radcliffe *et al.*, 1991). In the eastern and north central states, Colorado potato beetle is the key pest, and green peach aphid is a secondary pest. In the Maritimes, Maine, the intermountain states and the Pacific North-west, green peach aphid is the key pest.

In the eastern and north central states, Colorado potato beetle is the key potato pest and must be controlled from the time the crop emerges until harvest. Before neonicotinoid insecticides were available, repeated applications of foliar insecticides often resulted in insecticide-induced resurgence of green peach aphid. With the widespread use of systemic neonicotinoids at planting, resurgence of green peach aphid has not been encountered. Unfortunately, neonicotinoids tend to be less effective against potato leafhopper, *Empoasca fabae* (Harris), than were most of the foliar insecticides formerly used against Colorado potato beetle. In the north central states, growers using neonicotinoids may need to apply additional foliar insecticides for potato leafhopper control. Virtually any pyrethroid, organophosphate or carbamate insecticide will provide acceptable potato leafhopper control, but these products often induce aphid resurgence.

To avoid aphid resurgence, a recommended IRM strategy is to use insecticides below labelled rates for potato leafhopper control (Suranyi *et al.*, 1999). IRM usually cautions against this practice (http://www.irac-online.org), because the target insect is expected to evolve resistance faster. However, there are unique aspects of potato production that make this practice feasible for potato leafhopper. First, the other major pests in the system, Colorado potato beetle and green peach aphid, are already resistant to these products. Use of below-label rates aims to preserve predators and parasitoids that keep green peach aphid populations from flaring. Further, potato leafhopper has a broad host range and much of its habitat remains largely untreated in the region. Finally, potato leafhopper cannot survive the winter in the north central region and must migrate each spring from over-wintering sites along the Gulf of Mexico. Thus, even if resistance developed in one season and some individuals migrated back to overwintering sites, they would contribute little resistance to the large overwintering population, which is never treated with insecticides. It would appear that insecticide resistance in potato leafhopper will develop very slowly, and using below-label rates in potato to prevent flaring of green peach aphid populations outweighs the risks of exposing potato leafhopper to sub-lethal doses of insecticide.

In the Maritimes, Maine, the intermountain states and the Pacific North-west (and sometimes in the eastern and north central states), green peach aphid does reach treatable densities of 3–10/100 leaves for seed potatoes (Cancelado and Radcliffe, 1979; Hanafi *et al.*, 1989; Flanders *et al.*, 1991; Mowry, 2001) and 30–50/leaf for processing potatoes (Byrne and Bishop, 1979; Cancelado and Radcliffe, 1979; Shields *et al.*, 1984). Insecticides used for green peach aphid generally provide adequate control for potato leafhopper, so the IRM strategy is to avoid sequential use of products with the same mode of action. If neonicotinoids are applied at planting for early-season Colorado potato beetle control, the IRM recommendation for green peach aphid control is to follow with pymetrozine or methamidophos. Alternatively, if Colorado potato beetle is con-trolled with products other than neonicotinoids, neonicotinoid sprays can be used for aphid control. In some exceptional situations, the best option for some potato growers may be to follow an at-planting application of neonicotinoid with a foliar application of neonicotinoid. In this case, the IRM recommendation is to time the second neonicotinoid application early enough to avoid treating second-generation Colorado potato beetle. Here the strategy is to preserve sus-ceptibility in Colorado potato beetle, because there are few alternative chemicals for reliable beetle control.

In the Pacific North-west, two new pests have been identified recently: (i) the beet leafhopper, *Circulifer tenellus* (Baker) (Homoptera: Cicadellidae), which transmits the potato virescence agent (Lee *et al.*, 2004); and (ii) the potato tuber moth, *Phthorimaea operculella* (Zeller) (Lepidoptera: Gelechiidae), a late-season pest. If these new pests persist and require treatment over a broad ge-ographical area on an annual basis, IRM programmes in the Pacific North-west will need to be re-evaluated. An added concern for growers there is the propen-sity of two-spotted spider mites, *Tetranychus urticae* Koch (Acarina: Tetranychidae), to flare following multiple applications of pyrethroids and organophosphates, adding yet another pest to a system that has few alternative chemicals that provide reliable control.

Cotton bollworm, *Helicoverpa armigera*

It has been more than 20 years since the introduction of an IRM strategy in the Australian cotton industry. The strategy was implemented when pyrethroid resistance in *H. armigera*, the primary pest of cotton in Australia, resulted in unsatisfactory field control (Gunning *et al.*, 1984). Although voluntary, due to the severe threat to pest control posed by pyrethroid resistance, adoption of the strategy was close to 100% across all cotton-growing areas of eastern Australia, broadly encompassing nine distinct river valley systems from central Queensland through to southern New South Wales.

The approach taken to manage resistance involved consultation with representatives from all facets of the industry. These strategies incorporated an understanding of chemical use on all crops grown adjacent to cotton, in recognition that a coordinated approach across all cropping enterprises was necessary for effective management. Cooperation and coordination among all interested and affected parties were deemed essential to ensure full implementation of the strategy.

The strategy involved restricting pyrethroids to a window of use corresponding to the average life cycle of *H. armigera* (42 days), so that selection was limited to a single generation. This window of use, with the added restriction of a maximum of three applications during the window, was positioned to coincide with the important peak flowering/early boll set stage in cotton. This also corresponded to the peak sorghum flowering period where pyrethroids were used for sorghum midge control. Endosulfan was available for use prior to and during this window, and other insecticides with different modes of action, such as the organophosphates and carbamates, were used after the window. Additional guidelines included adding an ovicide to pyrethroid sprays under situations of high egg pressure, never following a suspected pyrethroid control failure with another pyrethroid application, and using at least three of the five different available modes of action when crops were receiving multiple sprays.

Non-chemical measures were also incorporated and outlined in the strategy to reduce the need for insecticide application. The IRM strategy was designed to fit with an IPM system, where the broad-spectrum pyrethroids are avoided early in the season to encourage the build-up of beneficial species to reduce *H. armigera* populations and replaced by the relatively softer chemicals, such as endosulfan and foliar *Bt*. Guidelines were provided regarding planting crops earlier to avoid high *H. armigera* populations later in the growing season. Thorough scouting and use of economic thresholds were recommended to reduce overspraying by optimizing the timing of spray applications. Non- or seldom-sprayed refuges and the incorporation of dryland sorghum, sunflowers and maize in both cotton- and non-cotton-growing regions were also an important source of dilution of pyrethroid resistance. While these refuges, along with other non-crop refuges such as native vegetation, were not specifically grown for this purpose, the increase in susceptibility at the commencement of each season was reliant on dilution by susceptible moths from these areas.

The voluntary strategy was well received by the industry, which faced the very real threat of losing the popular pyrethroids in a range of crops, and the frequency of resistance early in the growing season remained stable for some years (Forrester, 1989a). The restrictions on chemical use, particularly the pyrethroids and endosulfan, were strongly adhered to, as were the non-chemical components of the strategy. Various non-chemical measures became available and were taken up by the industry as IPM was developed. This included the incorporation of 'pupae busting' at the end of the season to break the resistance cycle between seasons. In 1989/1990, the strategy was modified after pyrethroid resistance at the end of the season increased across all valleys to levels higher than previously measured, and resistance in refuge habitats exceeded levels that could be readily diluted by the annual spring migration (Forrester, 1989b). In response, the pyrethroid window of use was limited to 35 days to further reduce selection. In addition, stricter guidelines regarding pyrethroid use were also added, including targeting only smaller larvae that could still be killed by pyrethroids.

In subsequent years, additional changes to the strategy were made as understanding of resistance improved, and the industry embraced alternative insect control methods and incorporated resistance management in sucking pests. In 1990/1991, the strategy was modified to include the synergist piperonyl butoxide (PBO) mixed with pyrethroids. Research had shown that the primary mechanism of nerve insensitivity common in 1983/1984 had been replaced by an enzyme-mediated resistance mechanism (Gunning *et al.*, 1991) that could be synergized by PBO. While this improved field control, resistance monitoring indicated that pyrethroid resistance was established firmly in the field and continued to increase (Forrester, 1992).

During the 1993/1994 season, the LepTon® Test Kit, which enabled relatively rapid differentiation between *H. armigera* and *Helicoverpa punctigera*, was introduced. Both species are severe pests, but *H. punctigera* remains susceptible to insecticides. As a consequence, the underlying logic of the IRM strategy shifted from limiting selection to a single generation to active avoidance of pyrethroid and endosulfan use on resistant *H. armigera* populations. Other restrictions on pyrethroid use were relaxed, removing the window of use, the maximum number of allowable applications, and the use restrictions on other crops. While pyrethroid restrictions were relaxed at this time, there is little doubt that the measures taken to manage resistance previously observed in the field prolonged their use until other strategies, such as that afforded by the LepTon® Test Kit, could be developed.

Modern resistance management remains a priority of the Australian cotton industry, but the focus has shifted from pyrethroid resistance to proactively managing the newer IPM-compatible chemistries. The pyrethroids are now restricted to use in the latter part of the growing season when disruption of beneficial insect natural enemies can be tolerated. Pyrethroid resistance is firmly established in *H. armigera* populations, and they are seldom applied without the addition of PBO or in a mixture with another insecticide with a different mode of action.

Possibly the greatest barrier to implementing effective resistance management is formulation of a strategy that effectively manages resistance but remains practical to the growers who must control variable pest complexes across a range of growing conditions. At its inception, compliance with the strategy was high, as the threat of pyrethroid resistance was readily observed in the field as unsatisfactory control. A proactive approach has since been adopted to prevent resistance development and, with the introduction of transgenic cotton and the registration of several new insecticides for *Helicoverpa* control, the threat of control failure from resistance is likely to be forgotten or unappreciated by many new to the industry.

Since 1996, formulation of the IRM strategy has been the responsibility of the Transgenic and Insect Management Strategy (TIMS) Committee, which is comprised of representatives from all aspects of the industry. This committee is supported by grower and consultant groups, and uses resistance monitoring results and other information to formulate effective IRM strategies. An annual resistance tour is undertaken by researchers to report resistance information and educate growers and consultants on the importance of resistance management. This tour provides an interface for researchers to receive feedback for consideration when formulating changes in the IRM strategy. The consequence has been a strategy that continues to evolve, reflecting a compromise between ideal IRM practices and realities associated with the availability of insecticidal options. Insecticides are no longer restricted to a window of use corresponding to a single *Helicoverpa* spp. generation, and generally applications have been extended to cover two generations with limits only on the maximum number of applications. The increased resistance risk is managed by the adoption of non-chemical control options, such as manipulation of beneficial populations, and the use of an increased number of insecticidal chemicals (12 including biologicals) with different modes of action. In contrast, the original pyrethroid IRM strategy had only four insecticidal modes of action available. Without this compromise, compliance is jeopardized, and effective IRM is unachievable.

5.3 Insect Resistance Management for Transgenic Crops

Bt cotton in Australia

As just discussed, insecticide resistance in *H. armigera* has been a consistent challenge for agricultural industries in Australia, particularly the cotton industry (Fitt, 1989, 1994; Forrester *et al.*, 1993). In 1996, with the introduction of the first generation of transgenic cottons expressing insecticidal proteins from *Bacillus thuringiensis* (*Bt*), this history of resistance provided a critically important context which guided a mature and measured approach by industry to the new technology. The first *Bt* cotton expressed the Cry1Ac protein from Monsanto under the trade-name INGARD® and was available in a suite of locally adapted varieties. More recently, varieties expressing two *Bt* proteins, Cry 1Ac and Cry2Ab, known as Bollgard II™, have been commercialized.

The possibility of resistance evolution to *Bt* cottons by *H. armigera* was a real concern. For this reason a pre-emptive resistance management strategy was implemented to accompany the commercial release of *Bt* varieties in 1996 (Roush *et al.*, 1998). Prior to that release, the cotton industry established the TIMS Committee to oversee the development and endorsement of strategies and interactions with regulatory bodies, and to provide critically important engagement of the cotton industry from the start of the transgenic era.

The Australian resistance management strategy for *Bt* cottons is based on the use of structured refuges to reduce selection pressure and maintain susceptible individuals in the population (Roush, 1996, 1997; Gould, 1998; Roush *et al.*, 1998) by taking advantage of the polyphagy and local mobility of *H. armigera* to utilize gene flow to counter selection for resistance in transgenic crops. In contrast, extensive natural refuges effectively nullify the resistance risk in *H. punctigera*. Indeed, *H. punctigera* provides an excellent natural example of the capacity of the refuge strategy to reduce resistance risk.

Key elements of the Australian strategy are the following (Monsanto, 2004):

1. Effective refuges on each farm growing *Bt* cotton, to reduce exposure and selection pressure.
2. Removal of volunteer *Bt* plants prior to sowing the next crop, to eliminate inadvertent selection pressure.
3. Defined planting window for *Bt* cotton to avoid late-planted crops that may be exposed to abundant *H. armigera* late in the growing season, to eliminate unnecessary selection pressure.
4. Mandatory cultivation of *Bt* cotton crops to destroy most overwintering pupae of *H. armigera*, to select against possible resistant pests.
5. Defined spray thresholds for *Helicoverpa* to ensure any survivors in the crops are controlled, to select against possible resistant pests.
6. Monitoring of *Bt* resistance levels in field populations.

Australian growers are able to choose from five different refuge options, each with a different area requirement determined by the relative productivity of the refuge (Fitt and Tann, 1996) compared with the 'control' refuge, which is unsprayed conventional cotton. Refuge requirements are somewhat larger than those required in other parts of the world (e.g. USA). For the differing options, the refuge sizes needed for each 100 ha of *Bt* cotton are: (i) 100 ha of sprayed conventional cotton (50% refuge); (ii) 10 ha of unsprayed conventional cotton (9%); (iii) 5 ha of unsprayed pigeon pea (4.8%); (iv) 15 ha of unsprayed sorghum (13%); or (v) 20 ha of unsprayed maize (16.7%). Refuge crops cannot be *Bt* crops or treated with *Bt* sprays, and must be in close proximity to the transgenic crops (within 2 km) to maximize the chances of effective mingling and mating among subpopulations (Dillon *et al.*, 1998).

Additional elements of conservatism in the IRM strategy were a phased introduction of INGARD® varieties and limitation of the area of INGARD® cotton to ≤ 30% of the total cotton area (70% refuge). The suggestion for an area limitation originated from cotton growers themselves and was implemented through the TIMS process as a means of further minimizing the risk of resistance to the Cry1Ac protein in preparation for the expected future release of two-gene combinations (*Cry1Ac/Cry2Ab*). In the first year, 1996, INGARD® varieties were grown on 30,000 ha representing about 8% of the total cotton area. In subsequent years, the area increased by 5% increments each year up to the 30% cap, which was reached in the 2000/2001 season.

Bollgard II™ varieties with two *Bt* genes were first approved for commercial use in 2002/2003 when they occupied only 5000 ha. The two-gene varieties provide much better efficacy and, hence, even greater reduction in pesticide requirement, but their main purpose is to provide much greater resilience against the risk of resistance (Roush, 1998). From the start, the Australian industry aimed for a rapid transition to Bollgard II™, with INGARD® varieties withdrawn after the 2003/2004 season. From 2004/2005 onwards, only Bollgard II™ varieties have been available, thereby removing the risk of mixed deployment of single- and two-gene varieties. As described in the next section, this policy is significantly different from that in the USA, where no mandated change to two-gene varieties is in place. Since 2004/2005 the area limitation on *Bt* cotton was also lifted, except for meeting the refuge requirements, and, in that season, *Bt* cotton occupied about 70% of the total cotton area. It is noteworthy that despite the reduced resistance risk with deployment of two *Bt* genes, all other components of the resistance management strategy remain unchanged, thereby maintaining a conservative resistance management approach.

The Australian IRM strategy also includes considerations to ensure compliance, auditing of compliance and monitoring of resistance. All aspects of the IRM plan for both INGARD® and Bollgard II™ cotton varieties are listed on the product label and are part of a single-use contract growers must sign with the technology provider (in this case, Monsanto) in order to purchase seed. Thus, the components of the IRM plan are legally binding on the grower. To support the contract and label, each farm growing *Bt* cotton must be audited three times each year to check on compliance with refuges, pesticide use and compulsory end-of-season cultivation. In some years, the technology licence fee for INGARD® cotton also included a rebate for growers who met the requirements of the three audits. That incentive is no longer provided with Bollgard II™ varieties, but compliance with the agreed IRM strategy has remained exceedingly high (Macarthur Agribusiness Report, 2004).

Prior to the commercial release of *Bt* cottons, a series of studies was completed over 4 years to establish the baseline susceptibilities of both *H. armigera* and *H. punctigera* to Cry1Ac toxin and some combinations of toxins. These studies demonstrated considerable geographic variation in tolerance to *Bt* toxins, but no evidence of resistance per se (Forrester, 1994). Since that time, a resistance monitoring programme has been in place to provide early indications of change in resistance gene frequencies in field populations. The monitoring system has evolved from reliance on a discriminating dose approach to the current programme, which involves both discriminating dose and an F_2 screen (Andow and Alstad, 1998) to provide more precise estimates of resistance gene frequencies for the *Bt* genes now deployed in Bollgard II™ cottons. To date, the monitoring programme has confirmed that the background frequency of Cry1Ac resistance is low and $<10^{-3}$ (Mahon *et al.*, 2004), and has shown no evidence of an increase in resistance to Cry1Ac after eight seasons of use in *Bt* cotton (Mahon *et al.*, 2004), despite reports of a low-level resistance strain (Gunning *et al.*, 2005). Likewise, there have been no changes detected in resistance to Cry2Ab after just two seasons of use, although the background frequency of Cry2Ab resistance is surprisingly high ($\sim 10^{-3}$) (Mahon *et al.*, 2004). The genetic basis and fitness characteristics of Cry1Ac and Cry2Ab resistance in Australian *H. armigera* are yet to be fully defined, but both appear to be recessive or semi-dominant and, at least in the case of Cry1Ac, associated with significant fitness penalties (Mahon, unpublished data).

Bt cotton in the USA

Bt cottons containing Cry1Ac received a provisional permit for commercialization in the USA from the EPA in 1995 (USEPA, 1998). They produce a high dose of the Cry1Ac protein against two of the target lepidopteran pests, *H. virescens* and *Pectinophora gossypiella* (Saunders) (Lepidoptera: Gelechiidae), but not against a third, *Helicoverpa zea* (Boddie) (Lepidoptera: Noctuidae) (Gould, 1998). A high dose is a level of *Bt* toxin expression that is enough to kill individuals heterozygous for resistance.

Because of the tremendous selection caused by continuous exposure to *Bt* toxin and the general concern for resistance evolution, the EPA appointed a scientific advisory panel (SAP) that recommended structured refuges be planted in conjunction with *Bt* cottons (USEPA-SAP, 1998). Unlike refuge options from Australia, these refuges consist only of non-*Bt* cotton. At present, three primary refuge options exist for *Bt* cotton in the USA: (i) 5% external unsprayed refuge; (ii) 20% external sprayed refuge; and (iii) 5% embedded refuge (USEPA, 2001). The 5% external unsprayed refuge must average at least 150 ft wide, be within half a mile of all associated *Bt* cotton fields, and may not be treated with any lepidopteran-active insecticides. The 20% external sprayed refuge must be within 1 mile of associated *Bt* cotton fields and may be treated with any insecticide other than foliar *Bt* products. The 5% embedded refuge must be planted within each *Bt* field or field unit, must average 150 ft in width (except for areas in Arizona, California, and New Mexico where *P. gossypiella* is the only lepidopteran pest of consequence and single-row refuges are allowed), and may be treated with the same insecticides and rates as the associated *Bt* cotton within a 24-h period.

The registrants of *Bt* cottons are also required by the EPA to survey grower compliance with regard to refuge requirements, maintain *Bt* resistance monitoring programmes and develop a remedial action plan in the case of resistance development. Grower compliance is assessed through surveys that are conducted by the registrants and independent parties. In addition, on-farm visits are utilized to verify compliance. Penalties for grower non-compliance include: (i) *Bt* sales restrictions at a county, state, regional or national level; (ii) sales prohibitions to specific growers; (iii) registrant fines; and (iv) increased refuge requirements for specific growers (USEPA, 2001). One registrant has reported compliance levels with EPA ranging from 87 to 99% during 1996–2004, although this has not been independently confirmed.

Registrants must also monitor for insect resistance to *Bt* toxins as an important early warning sign to field-level resistance. Currently, the US Department of Agriculture (USDA)/Agricultural Research Service (ARS)/Southern Insect Management Research Unit (SIMRU) and University of Arkansas maintain a *Bt* resistance monitoring programme for *H. virescens* and *H. zea*, whereas the

University of Arizona Extension Arthropod Resistance Management Laboratory (EARML) monitors *Bt* resistance in *P. gossypiella*. To date, susceptibility of *H. virescens* and *P. gossypiella* to *Bt* proteins has not decreased in field populations (Tabashnik *et al.*, 2003; Blanco *et al.*, 2004). Some degree of increased tolerance to Cry1Ac was found in *H. zea* populations from 1996 to 1999 (Summerford *et al.*, 1999). However, increases in the frequency of *Bt* resistance alleles have not been observed in *H. zea* populations (Blanco *et al.*, 2004; Jackson *et al.*, 2006), and tolerance levels to *Bt* proteins have been variable (Luttrell *et al.*, 1999; Ali *et al.*, 2006).

A further requirement for the registrant of *Bt* cotton is to develop a remedial action plan in the case of putative resistance. The remedial action plan includes: (i) monitoring for early detection of resistance through commercial field evaluations (conducted by consultants, producers, and research/extension personnel), sentinel plot programmes and monitoring programmes (USDA/ARS/ SIMRU, University of Arkansas, and University of Arizona/EARML); (ii) identifying putative resistance; (iii) confirming resistance; and (iv) implementing mitigation actions. Mitigation actions include communicating resistance problems to stakeholders, implementing alternative population suppression tactics, increasing monitoring in the affected areas for changes in resistance allele frequencies, modifying refuges in the affected areas, ceasing sales in the affected and bordering counties, and replacing Cry1Ac-producing varieties with other types of transgenic cotton.

Scientists have demonstrated the potential for *Bt* resistance evolution on numerous occasions, as various strains of *H. zea*, *H. virescens* and *P. gossypiella* have developed high levels of resistance to several *Bt* toxins in the laboratory (Gould *et al.*, 1992, 1995; Liu *et al.*, 1999, 2001a,b; Luttrell *et al.*, 1999; Tabashnik *et al.*, 2000, 2002a,b; Burd *et al.*, 2003; Morin *et al.*, 2003; Jackson *et al.*, 2006). Furthermore, experiments have demonstrated that *H. zea* and *P. gossypiella* violate key assumptions of the high-dose/refuge strategy (low initial *Bt* resistance allele frequency or recessive inheritance of resistance to *Bt* toxins) (Tabashnik *et al.*, 2000; Burd *et al.*, 2003). Thus because of widespread use of *Bt* cottons, resistance might develop in certain situations despite the implementation of current IRM strategies (Tabashnik *et al.*, 2003).

To date, resistance failures to *Bt* cotton have not been documented at the field level, and increases in the frequency of *Bt* resistance alleles in general target pest populations have not been detected (Tabashnik *et al.*, 2000; Blanco *et al.*, 2004; Jackson *et al.*, 2006), although increased tolerance to Cry1Ac in *H. zea* was reported over a 3-year period (Summerford *et al.*, 1999). However, it is not yet possible to conclude that the current IRM strategy has delayed resistance evolution. The strategy may have been responsible for the delay due to one or more of the following factors: fitness costs experienced by resistant individuals, low initial frequency of resistant alleles, a high dose of toxin delivered by the plant or adequate production of *Bt*-susceptible individuals from refuge cottons (Bates *et al.*, 2005). It is also possible that non-cotton refuges, which are not a part of the strategy, have been responsible for the delay in some parts of the USA. Stable carbon isotope and adult emergence studies have shown that C_4 plant hosts, most likely maize, and other C_3 plants, like groundnut and soybean, probably serve as important refuges for *H. zea* throughout much of south-eastern USA (Gould *et al.*, 2002; Jackson *et al.*, 2004). In addition in North Carolina, tobacco may be an alternative crop host for *H. virescens* (Abney *et al.*, 2004), and preliminary results from gossypol-detection analyses suggest that non-cotton crops and/or wild hosts play an integral role in *H. virescens* refuge contributions (R. Jackson, unpublished data).

Barriers to the implementation of and compliance with the present IRM strategy are heightened by this uncertainty. One deterrent is the cost associated with maintaining a refuge. Economic analyses suggested that reduced refuge requirements would improve grower profitability and that sprayed refuges were more cost effective than unsprayed refuges (Livingston *et al.*, 2004). Also, cottons that produce the Cry1Ac endotoxin have been shown to exhibit increased profit margins of $16 to almost $173/acre over non-*Bt* cottons managed with pesticides (Marra *et al.*, 2002). Another barrier is the lack of knowledge about resistance management in the grower community. As long as growers do not understand why they must forego increased profits on at least 5% of their cotton acreage, then grower compliance may never reach 100%.

To maximize the environmental benefits of *Bt* technologies and ensure their sustainability, IPM approaches should be utilized in conjunction with transgenic crops (Bates *et al.*, 2005). While second-generation pyramided *Bt* cottons have already been commercialized, a database is being developed by academia and industry that may support the reassessment of current IRM strategies for *Bt* cotton by the EPA. Alternative IRM strategies could be proposed in portions of the cotton-growing region to include options for alternative crop hosts, similar to refuge options available in Australia.

Bt maize

In the USA, four *cry* transgene events are presently approved for commercial use. Two of these, Bt-11 and Mon810, express Cry1Ab and are active against the main lepidopteran borer pests, including *Ostrinia nubilalis* (Hübner) and *Diatraea grandiosella* (Dyar) (Lepidoptera: Crambidae). Neither of these species has a history of resistance to insecticides, although little insecticide has been applied for their control historically. Bt-11 is also used in sweetcorn. A third event, TC1507, expresses Cry1F which is also active against the main lepidopteran borer pests. These three events are also labelled for control of *H. zea* and *Spodoptera frugiperda* (J.E. Smith) (Lepidoptera: Noctuidae), and some are labelled for control of *Diatraea crambidoides* (Grote), *Diatraea saccharalis* (F.) (Lepidoptera: Crambidae), and *Papaipema nebris* (Guenee) (Lepidoptera: Noctuidae). All are known or likely to express at a high dose against the crambid borer pests and to express at a low dose against all of the noctuid pests. IRM for these events has focused on the widespread pests, *O. nubilalis*, *D. grandiosella* and *H. zea*. The fourth event, Mon863, which expresses Cry3Bb1, is targeted against several corn rootworm species, *Diabrotica virgifera virgifera* (LeConte), *Diabrotica barberi* (Smith and Lawrence) and *Diabrotica virgifera zea* (Krysan and Smith) (Coleoptera: Chrysomelidae). This event does not express at a high dose against any of these pests. IRM has focused on *D. v. virgifera*, which has a history of resistance to insecticides. In addition, a combination of Mon810 and Mon863, and a modified T1507 event have been registered, and several other maize events are under consideration. Two other Cry1Ab events had been registered (event 176 and Mon801), but these registrations were cancelled.

In 1995, at the time of the initial conditional registrations of *Bt* maize, there was no scientific consensus on IRM needed to delay resistance in the target pests. At that time, suggested values for refuge size ranged from 0 to 50% of non-*Bt* maize or other host plants per farm. By the end of 1995, IRM became a required component for registration. Early in 1997, the USDA regional research committee NC-205 reviewed model results and information on the ecology of *O. nubilalis* and suggested to registrants and the EPA that a 20–25% refuge was needed near all *Bt* maize fields with lepidopteran-active Cry toxins (Anon., 1998). Research results supporting this recommendation were published in the ensuing years (Onstad and Gould, 1998; Hunt *et al.*, 2001; Showers *et al.*, 2001; Bourguet *et al.*, 2003). One of the key results was a bioeconomic model suggesting that a 20% refuge would be nearly optimal for growers who consider the trade-off between the immediate costs of the refuge and delayed costs of resistance failures (Hurley *et al.*, 2001).

Canada required a 20% refuge within 0.5 miles (~800 m) of *Bt* maize in 1998, and, during 1999, a coalition of *Bt* maize registrants (working with the National Corn Growers Association) and the Agricultural Biotechnology Stewardship Technical Committee (a group of cooperating biotechnology companies) approached the EPA with a uniform IRM plan for their lepidopteran-active products (20% refuge within 0.5 miles of *Bt* maize). Starting in 2000, the following plan was implemented to manage resistance in *O. nubilalis*, *D. grandiosella* and *H. zea* (USEPA, 2001).

1. All growers must plant 20% non-*Bt* field maize refuge within 0.5 miles of their *Bt* maize; the refuge can be sprayed with any insecticide except a *Bt* insecticide if economic thresholds are exceeded. The refuge reduces selection pressure on the pest, and, when *Bt* maize is high-dose, it produces susceptible individuals that can mingle and mate with resistant ones from *Bt* fields, further delaying resistance evolution. If insecticides have historically been used to control the target pests, the refuges should be within 0.25 miles of the *Bt* fields, so susceptible adults can more readily mingle and mate with resistant ones from *Bt* fields. In cotton-growing regions, a 50% non-*Bt* field maize refuge within 0.5 miles of *Bt* field maize is required to mitigate the resistance risk in *H. zea*, which will be selected for resistance by both *Bt* maize and *Bt* cotton. This refuge can be sprayed with any insecticide except a *Bt* insecticide if economic thresholds are exceeded. Three refuge planting options are provided.

2. A compliance assurance programme (CAP) to promote grower compliance to the IRM requirements, including grower educational materials that raise awareness of IRM and clarify IRM requirements so that all growers understand how to implement these requirements, annual contracts to obligate growers to follow the requirements, training of seed company personnel, an annual survey of compliance conducted by a third party, a mechanism for handling tips and complaints about non-compliance and an auditing system to visit farms and verify compliance. Growers who do not comply with the IRM requirements in two consecutive years will be denied access to *Bt* maize the third year. If regions of non-compliance are identified, the CAP will be adjusted accordingly.

3. Monitoring for resistance and resistance failures, to include annual monitoring for the development of resistance in the target pests, a toll-free customer service telephone line so customers can report unexpected levels of damage from the target pests and tests to verify these reports. Annual monitoring should focus on the geographic areas with the highest use of *Bt* maize.

4. Remedial action if there is suspected resistance. If resistance occurs, sales and distribution will be stopped county-wide, and a local remedial action plan must be developed, approved and implemented. The initial remedial action will be to stop sales of *Bt* maize products.

5. Reporting the sales quantities of *Bt* maize on a county-by-county basis, so that regions with high use of *Bt* maize can be identified.

Refuge maize should have similar varietal characteristics (e.g. growth, maturity and yield) and crop management (e.g. cultivation, fertility, irrigation, weed management and planting date) to the *Bt* maize. Refuges may be planted in separate fields, as blocks within fields (i.e. in the headlands, along one edge or in the middle of the field) or in strips across the field. When planting strips, growers are encouraged to plant multiple non-*Bt* maize rows whenever possible (preferably six or more rows). Temporal refuges, non-maize refuges and seed mixtures are not recommended.

For lepidopteran-active *Bt* sweetcorn, the target species differ from field maize, and are *O. nubilalis*, *H. zea* and *S. frugiperda*. Refuges are not required because sweetcorn harvesting occurs before ear-feeding insects mature and reproduce. However, other aspects of resistance management are required, including the following:

1. Mandatory destruction of *Bt* sweetcorn stalks immediately after or within 1 month of harvest. Stalk destruction is intended to reduce the possibility of any insects, including resistant insects, surviving to the next generation.

2. Restrictions from using *Bt* microbial insecticides on *Bt* sweet corn.

3. No sale of seed to small-scale roadside and home growers, so that compliance with the IRM requirements will be more readily assured.

4. A grower education programme to improve compliance with the IRM requirements, including materials that raise awareness of IRM and clarify IRM requirements, so that all growers understand how to implement these requirements.

5. Monitoring for resistance and resistance failures, to include annual monitoring for the development of resistance in the target pests, a toll-free customer service telephone line so customers can report unexpected levels of damage from the target pests, and tests to verify these reports.

6. Remedial action if resistance is confirmed, including notification of affected parties, recommendation of alternative controls of target pests, required incorporation of crop residues in the soil immediately after harvest, intensified monitoring to establish the geographic extent of resistance, development and implementation of a modified IRM plan, and implementation of a structured refuge. If these fail to mitigate resistance, sales will be stopped voluntarily.

IRM for the rootworm-active event integrates IRM for rootworms and Lepidoptera, so that the combined product, both Cry1Ab and Cry3Bb1, can be used with effective IRM (USEPA, 2005). The IRM requirements are largely the same as for the lepidopteran-active field maize events, with some differences in refuge requirements. Two refuge options are available. The first is a refuge that is effective for both the lepidopteran and rootworm target pests ('common refuge'). This is planted with a variety that does not have any of the *Bt cry* genes, is at least 20% of the grower's maize area, and is in a block adjacent to the combined *Bt* product, in border rows, or in strips in the same field as the combined *Bt* product. If in border rows or strips, it must be at least six rows wide, and preferably 12 rows wide. The common refuge can be treated with any seed- or soil-applied insecticide and any non-*Bt* foliar insecticide except when adult rootworms are present. If adult rootworms are present, both the refuge and the *Bt* field must receive the same foliar application. The second option is separate refuges for Lepidoptera and rootworms. The lepidopteran refuge must meet the requirements above for the lepidopteran-active events. The rootworm refuge must be a non-Cry3Bb1 variety, must be at least 20% of the grower's maize area, and otherwise must meet the requirement of the common refuge. In cotton-growing areas, the common refuge is 50% of the grower's maize area and otherwise must be managed like the maize-only common refuge. If separate refuges are used, the lepidopteran refuge must meet the requirements above for the lepidopteran-active events (50% refuge), and the rootworm refuge must meet the requirements of the maize-only separate refuge.

Compliance with the lepidopteran IRM strategy has been estimated by various surveys. Range-wide, an industry-sponsored telephone survey of 501 farmers with >90 ha of maize in 2000 found that 87% planted ≥ 20% refuge, and 82% planted refuges within the required 800 m of the *Bt* crop (USEPA, 2001). In combination, 71% of growers were in compliance with the IRM refuge requirements. However, using 2002 data from the USDA National Agricultural Statistics Service, Jaffe (2003) reported that only 79% of farmers in ten Midwest states complied with the 20% refuge size requirement alone, and this was especially low among farms with <90 ha of maize. Industry repeated the telephone survey between 2000 and 2004, and found compliance with the 20% size requirement had increased from 87 to 91% and compliance with the 800-m distance requirement had increased from 82 to 96% (ABSTC, 2005). A mail survey of growers in Minnesota and Wisconsin for the 2004 growing season, however, found 24–28% non-compliance (Goldberger *et al.*, 2005). Clearly there is considerable variation in estimated rates of compliance. Little research exists that explores the resistance risks associated with such levels of non-compliance.

Considerable effort has gone into addressing the need for implementing an effective IRM strategy for *Bt* maize. Despite this effort, it must be acknowledged that there is little way of confirming that the strategy has been effective. *Bt* maize was first commercially available during 1995, but its use is highly variable geographically. Even in areas of high use, it has exceeded 50% for only 5 years. Hence pests have not been under high selection pressure for very long. Models of resistance evolution suggest that the present strategy should delay the onset of resistance by at least 15–20 years even when use rates are at the maximum 80% (Onstad and Gould, 1998). Thus, it is too early to tell if resistance evolution has been substantially delayed by the present IRM strategy.

5.4 Conclusions

During the past decade, there has been increased development of pre-emptive IRM strategies (Denholm and Rowland, 1992) for both insecticides and transgenic insecticidal crops. This occurred first in the southern USA (Plapp *et al.*, 1990) and Australia with pyrethroid IRM in cotton during the 1980s, which led to the TIMS Committee in Australia guiding IRM for *Bt* cotton. In the USA, pre-emptive IRM for insecticides and transgenic crops has developed in parallel, continuing in the mid-1990s with the IRAC and several USEPA-SAPs.

All IRM strategies depend first and foremost on methods to reduce the selection pressure of the insecticide or insecticidal crop on the target pests. However, the methods used differ substantially for insecticides versus insecticidal crops (Table 5.2). For insecticides, rotation of product mode of action and reducing the need for insecticide application through effective alternative control practices in an IPM system have essential roles, while refuges, when present, are typically unplanned and unstructured. For insecticidal crops, planned, structured refuges are sometimes the sole method for IRM, although practices that minimize the need to use the insecticidal crop and methods that select against resistance are also components of some of the IRM strategies.

Historical and ecological factors may explain some of these differences. Although refuge strategies have been proposed for some time (Comins, 1977; Georghiou and Taylor, 1977), they have been considered too complicated for insecticide IRM. In particular, when insecticides are used with scouting and economic thresholds, it is difficult to convince a grower to leave some of the crop as an unsprayed refuge. Instead, it is easier to rotate modes of action and to minimize the need for insecticide sprays, as both of these IRM tactics are consistent with grower goals to increase profit, minimize risk and reduce management time. In contrast, when pest control is pre-emptive (insecticidal crops, insecticides applied at planting), refuges may be implemented as a part of planting, and planned-for areas may be less likely to suffer economic losses. In addition, while average expected pest losses to the refuge can be calculated, in any given year these losses may or may not be incurred. In this context, refuges do not necessarily reduce profits, especially when longer timeframes are considered (Hurley *et al.*, 2001), and they do not necessarily increase risk.

IRM for *Bt* cotton has developed very differently in Australia and the USA (Table 5.2). Australia allows non-cotton crop refuges, requires larger refuge populations, specifies a planting window, requires use of economic thresholds to manage pests on refuges, and requires control of volunteer plants and

Table 5.2. Summary of Insect Resistance Management (IRM) strategies used in the field.

	IRM responsive/ pre-emptive	Refuges	Pest management requirements to avoid exposure to product	Temporal variation in selection	Counter-selection	Monitoring resistance/ failures	Compliance and audits	Remedial action plan
Myzus persicae	Responsive	Natural Unstructured	Prevent resurgence, other modes of action and reduced rates	Rotation of products	None	None	None	None
Leptinotarsa decemlineata	Responsive	No size None	Crop rotation	Rotation of products	None	None	None	None
Helicoverpa armigera	Responsive	Dryland crops Unstructured	Window of use Limit no. of applications Other modes of action	Rotation of products	Pupae busting	Resistance	None	None
Bt cotton, Australia	Pre-emptive	No size Crops Structured (2 km)	Planting window Volunteer removal	None	Residue destruction	Resistance Failures: via consultant network	Audits 3 times per year	Modify IRM plan
Bt cotton, USA	Pre-emptive	Size (≤ 50%) Cotton Structured Size (≤ 20%)	Spray thresholds None	None	None	Both required	Surveys, some growers audited	Stop sales, modify IRM plan
Bt field maize, lepidopteran-active	Pre-emptive	Maize Structured (800 m) Size (≥ 20%)	None	None	None	Both required	CAP, some growers audited	Stop sales, modify IRM plan
Bt sweetcorn, Cry1Ab	Pre-emptive	None	None	None	Residue destruction	Both required	CAP	Modify IRM plan
Bt maize, Cry3Bb1	Pre-emptive	Maize Structured (800 m) Size (≥ 20%)	None	None	None	Both required	CAP	Stop sales, modify IRM plan

CAP, Compliance Assurance Programme.

destruction of crop residues. The USA requires cotton refuges, allowing only 5% refuge area or non-structured non-cotton refuges, and does not require any other IRM measure. The Australian requirements are more risk-averse than the US requirements. These differences are in part due to the history of resistance failures in Australia that have sensitized growers to the problem. Probably more significant, however, is the TIMS process used in Australia, which involves the growers in the development of the IRM strategy. By doing this, growers are informed of the need for IRM, can influence the development of IRM so that it is consistent with their production goals, and are prepared to implement and comply with the requirements. In contrast, the USA uses a regulatory process that focuses on the registrant and limits grower inputs to consultations, because the growers are not the product registrants. This means that growers are less invested in the IRM strategy and must be convinced of the need and benefits after it has been proposed.

Potato production poses some unique challenges and opportunities with its high reliance on insecticides, multiple pests which are resistant to several chemicals having different modes of action, and few natural refuges for the pests. These concerns have been well recognized by all stakeholders involved in potato production in the USA, and growers have been actively involved in developing, promoting and implementing IRM for neonicotinoids and fungicides (IRAC, 2005; NPC, 2005). As with any crop production system, the entire pest complex must be considered, and the unique biology each pest presents must be taken into account in developing an effective IRM programme. In potato, IRM for one species can result in ineffective control of another pest, increasing the cost of pest control. The challenge to growers and their advisors is to maintain adequate insect control while practising sound IRM. The consolidation of agrichemical companies worldwide, the slowing development of new insecticides and the alarming rate at which insecticide resistance continues to evolve have converged to make a persuasive argument for more rapid development and promotion of sound IRM programmes. Crop professionals will need to be ever vigilant to keep the few effective insecticides available that still provide reliable control of green peach aphid and Colorado potato beetle, while at the same time not hastening development of resistance in other potato pests.

It is not yet possible to know for certain how effective IRM has been at delaying the onset of resistance for any of the cases in this chapter (Tabashnik *et al.*, 2003). However, the pyrethroid and *Bt* cotton IRM strategies in Australia surely delayed the rate of resistance evolution. Based on the rate of resistance evolution in cotton pests with no IRM strategy, it is likely that resistance would have occurred faster without the pyrethroid IRM strategy; although without further study it cannot be determined how much faster resistance would have occurred. Similarly, present knowledge about the commonness and inheritance of resistance to single-gene *Bt* cotton in *H. armigera* (Akhurst *et al.*, 2003) would indicate that some resistance failures would have been likely had its widespread use occurred in Australia. In addition, this would have jeopardized the IRM strategy for the two-gene *Bt* cotton.

In general, resistance problems have been greatest on those crops where insecticides have been used as the primary method of pest control. In these circumstances, there are relatively few alternative control measures, such as cultural control, host plant resistance or biological control. While strict cosmetic standards require high insecticide use for fruits, nuts and some vegetables, this is only part of the reason for high insecticide use. To improve IRM for these crops, it will be necessary to develop effective alternative control measures that can reduce the need for insecticides and insecticidal crops. In addition, there is a need to understand better how growers evaluate the effect of IRM on net profit, risk and management costs within their farm operations. Without this understanding, it will be difficult to anticipate and ensure grower compliance with IRM requirements. While there are many research needs to improve IRM worldwide, we believe that these two knowledge gaps are the most substantial barriers for implementing IRM for the pests with the most extensive history of resistance.

References

Abney, M.R., Sorenson, C.E. and Bradley, J.R. Jr (2004) Alternate crop hosts as resistance management refuges for tobacco budworm in NC. In: *Proceedings of the Beltwide Cotton Conference, San Antonio, Texas, 5–9 January 2004*. National Cotton Council of America, Memphis, Tennessee, pp. 1413–1416.

ABSTC (2005) Insect Resistance Management Grower Survey for Corn Borer-Resistant *Bt* Field Corn: 2004 Growing Season. http://www.pioneer.com/CM Root/Pioneer/biotech/irm/survey.pdf (accessed December 2007).

Akhurst, R.J., James, W., Bird, L.J. and Beard, C. (2003) Resistance to the Cry1Ac δ–Endotoxin of *Bacillus thuringiensis* in the Cotton Bollworm, *Helicoverpa armigera* (Lepidoptera: Noctuidae). *Journal of Economic Entomology* 96, 1290–1299.

Ali, M.I., Luttrell, R.G. and Young, S.Y. III (2006) Susceptibilities of *Helicoverpa zea* and *Heliothis virescens* (Lepidoptera: Noctuidae) populations to Cry1Ac insecticidal protein. *Journal of Economic Entomology* 99, 164–175.

Andow, D.A. and Alstad, D.N. (1998) The F_2 screen for rare resistance alleles. *Journal of Economic Entomology* 91, 572–578.

Anon. (1998) Supplement to: *Bt* corn & European corn borer: long-term success through resistance management, NCR-602. http://www.entomology.umn.edu/ecb/nc205doc.htm (accessed November 2004).

Anthon, E.W. (1955) Evidence for green peach aphid resistance to organophosphate insecticides. *Journal of Economic Entomology* 48, 56–57.

Bates, S.L., Zhao, J.-Z., Roush, R.T. and Shelton, A.M. (2005) Insect resistance management in GM crops: past, present, and future. *Nature Biotechnology* 23, 57–62.

Bishop, B.A. and Grafius, E. (1996) Insecticide resistance in the Colorado potato beetle. In: Jolivet, P.H.A. and Cox, M.L. (eds) *Chrysomelidae Biology*, Vol. 1. SBP Academic Publishing, Amsterdam, pp. 355–377.

Bishop, B., Grafius, E., Byrne, A., Pett, W. and Bramble, E. (2003) *Potato Insect Biology and Management. 2002 Research Report*. Michigan Potato Research Report No. 34. Michigan Agricultural Experiment Station, East Lansing, Michigan, pp. 113–130.

Blackman, R.L. and Eastop, V.F. (2000) *Aphids on the World's Crops: An Identification and Information Guide*, 2nd edn. Wiley, Chichester, UK.

Blanco, C.A., Adams, L.C., Gore, J., Hardee, D., Mullen, M., Bradley, J.R., Van Duyn, J., Ellsworth, P., Greene, J., Johnson, D., Luttrell, R., Studebaker, G., Herbert, A., Karner, M., Leonard, R., Lewis, B., Lopez, J.D. Jr, Parker, D., Williams, M., Parker, R.D., Roof, M.,

Sprenkel, R., Stewart, S., Weeks, J.R., Carroll, S., Parajulee, M., Roberts, P. and Ruberson, J. (2004) *Bacillus thuringiensis* monitoring program for tobacco budworm and bollworm in 2003. In: *Proceedings of the Beltwide Cotton Conference, San Antonio, Texas, 5–9 January 2004*. National Cotton Council of America, Memphis, Tennessee, pp. 1327–1331.

Bourguet, D., Chaufaux, J., Séguin M., Buisson, C., Hinton, J.L., Stodola, T.J., Porter, P., Cronholm, G., Buschman, L.L. and Andow, D.A. (2003) Frequency of alleles conferring resistance to *Bt* maize in French and US corn belt populations of the European corn borer, *Ostrinia nubilalis*. *Theory of Applied Genetics* 106, 1225–1233.

Brown, T.M. (1981) Countermeasures for insecticide resistance. *Bulletin of the Entomological Society of America* 27, 198–201.

Burd, A.D., Gould, F., Bradley, J.R., Van Duyn, J.W. and Moar, W.J. (2003) Estimated frequency of nonrecessive *Bt* resistance genes in bollworm, *Helicoverpa zea* (Boddie) (Lepidoptera: Noctuidae) in eastern North Carolina. *Journal of Economic Entomology* 96, 137–142.

Byrne, A., Grafius, E., Bishop, B. and Pett, W. (2004) Susceptibility of Colorado potato beetle populations to imidacloprid and thiamethoxam. *Arthropod Management Tests* 29, L12.

Byrne, D.N. and Bishop, G.W. (1979) Relationship of green peach aphid numbers to spread of potato leaf roll virus in southern Idaho. *Journal of Economic Entomology* 72, 809–811.

Cancelado, R.E. and Radcliffe, E.B. (1979) Action thresholds for green peach aphid on potatoes in Minnesota. *Journal of Economic Entomology* 72, 606–609.

Comins, H.N. (1977) The development of insecticide resistance in the presence of migration. *Journal of Theoretical Biology* 64, 177–197.

Davis, J.A., Radcliffe, E.B. and Ragsdale, D.W. (2003) Control of aphids on potatoes using foliar insecticides, 2002. E48. http://www.entsoc.org/protected/amt/amt28/text/E48.htm

Denholm, I. and Rowland, M.W. (1992) Tactics for managing pesticide resistance in arthropods: theory and practice. *Annual Review of Entomology* 37, 91–112.

Dennehy, T.J. (1987) Decision-making for managing pest resistance to pesticides. In: Ford, M.G., Holloman, D.W., Khambay, B.P.S. and Sawicki, R.M. (eds) *Combating Resistance to Xenobiotics: Biological and Chemical Approaches*. Ellis Horwood, Chichester, UK, pp. 118–126.

Devonshire, A.L., Field, L.M., Foster, S.P., Moores, G.D., Williamson, M.S. and Blackman, R.L. (1998) The evolution of insecticide resistance in the peach-potato aphid, *Myzus persicae*. *Philosophical Transactions of the Royal Society of London, Series B* 353, 1677–1684.

Dillon, M.L., Fitt, G.P. and Zalucki, M.P. (1998) How should refugia be placed upon the landscape? A modelling study considering pest movement and behaviour. In: Zalucki, M.P., Drew, R.A.I. and White, G.G. (eds) *Pest Management – Future Challenges*. University of Queensland Press, Brisbane, Australia, pp. 179–189.

Dively, G., Follett, P., Linduska, J. and Roderick, G. (1998) Use of imidacloprid-treated row mixtures for Colorado potato beetle (Coleoptera: Chrysomelidae) management. *Journal of Economic Entomology* 91, 376–387.

Fitt, G.P. (1989) The ecology of *Heliothis* species in relation to agroecosystems. *Annual Review of Entomology* 34, 17–52.

Fitt, G.P. (1994) Cotton pest management: Part 3. An Australian perspective. *Annual Review of Entomology* 39, 543–562.

Fitt, G.P. and Tann, C. (1996) Quantifying the value of refuges for resistance management of transgenic *Bt* cotton. In: Swallow, D. (ed.) *Proceedings of the Australian Cotton Conference*, pp. 77–83.

Flanders, K.L., Radcliffe, E.B. and Ragsdale, D.W. (1991) Potato leafroll virus spread in relation to densities of green peach aphid (Homoptera: Aphididae): implications for management thresholds for Minnesota seed potatoes. *Journal of Economic Entomology* 84, 1028–1036.

Forrester, N. (1989a) Update insecticide resistance levels. *The Australian Cottongrower* (Feb–Apr), 30–34.

Forrester, N. (1989b) Six years experience with the strategy. *The Australian Cottongrower* (Aug–Oct), 62–64.

Forrester, N. (1992) *Heliothis* resistance research update. *The Australian Cottongrower* (May–Jun), 12–18.

Forrester, N. (1994) Use of *Bt* in integrated control, especially on cotton pests. *Agriculture, Ecosystems and Environment* 49, 77–83.

Forrester, N.W., Cahill, M., Bird, L.J. and Layland, J.K. (1993) Management of pyrethroid and endosulfan resistance in *Helicoverpa armigera* (Lepidoptera: Noctuidae) in Australia. *Bulletin of Entomological Research Supplement* No. 1.

Georghiou, G.P. and Taylor, C.E. (1977) Genetic and biological influences in the evolution of insecticide resistance. *Journal of Economic Entomology* 70, 319–323.

Goldberger, J., Merrill, J. and Hurley, T. (2005) *Bt* corn farmer compliance with insect resistance management requirements in Minnesota and Wisconsin. *AgBioForum*, 8(2,3), 12. http://www. agbioforum.org/v8n23/v8n23a12-hurley.htm (accessed December 2007).

Gould, F. (1998) Sustainability of transgenic insecticidal cultivars: integrating pest genetics and ecology. *Annual Review of Entomology* 43, 701–726.

Gould, F., Martinez-Ramirez, A., Anderson, A., Ferre, J., Silva, F.J. and Moar, W.J. (1992) Broad-spectrum resistance to *Bacillus thuringiensis* toxins in *Heliothis virescens*. *Proceedings of the National Academy of Sciences USA* 89, 7986–7990.

Gould, F., Anderson, A., Reynolds, A., Bumgarner, L. and Moar, W. (1995) Selection and genetic analysis of a *Heliothis virescens* (Lepidoptera: Noctuidae) strain with high levels of resistance to *Bacillus thuringiensis* toxins. *Journal of Economic Entomology* 88, 1545–1559.

Gould, F., Blair, N., Reid, M., Rennie, T.L., Lopez, J. and Micinski, S. (2002) *Bacillus thuringiensis*-toxin resistance management: stable isotope assessment of alternate host use by *Helicoverpa zea*. *Proceedings of the National Academy of Sciences USA* 99, 16581–16586.

Grafius, E. (1997) Economic impact of insecticide resistance in the Colorado potato beetle (Coleoptera: Chrysomelidae) on the Michigan potato industry. *Journal of Economic Entomology* 90, 1144–1151.

Grafius, E., Bishop, B., Pett, W., Byrne, A. and Bramble, E. (2004) *Potato Insect Biology and Management*. Michigan Potato Research Report No. 35. Michigan Agricultural Experiment Station, East Lansing, Michigan, pp. 96–109.

Grafius, E., Pett, W., Bishop, B., Byrne, A. and Bramble, E. (2005) *Potato Insect Biology and Management*. Michigan Potato Research Report No. 36. Michigan Agricultural Experiment Station, East Lansing, Michigan (in press).

Guenthner, J.F., Wiese, M.V., Pavlista, A.D., Sieczka, J.B. and Wyman, J. (1999) Assessment of pesticide use in the US potato industry. *American Journal of Potato Research* 76, 25–29.

Gunning, R.V., Easton, C.S., Greenup, L.R. and Edge, V.E. (1984) Pyrethroid resistance in *Heliothis armigera* (Hübner) (Lepidoptera: Noctuidae) in Australia. *Journal of Economic Entomology* 77, 1283–1287.

Gunning, R.V., Easton, C.S., Balfe, M.E. and Ferris, I.G. (1991) Pyrethroid resistance mechanisms in Australian *Helicoverpa armigera*. *Pesticide Science* 33, 473–490.

Gunning, R.V., Dang, H., Kemp, F., Nicholson, I. and Moores, G. (2005) New resistance mechanism in *Helicoverpa armigera* threatens transgenic crops expressing *Bacillus thuringiensis* Cry IAc toxin. *Applied and Environmental Microbiology* 71, 2558–2563.

Hanafi, A., Radcliffe, E.B. and Ragsdale, D.W. (1989) Spread and control of potato leafroll virus in Minnesota. *Journal of Economic Entomology* 82, 1201–1206.

Hare, J.D. (1990) Ecology and management of the Colorado potato beetle. *Annual Review of Entomology* 35, 81–100.

Harrewijn, P. and Kayser, H. (1997) Pymetrozine, a fast-acting and selective inhibitor of aphid feeding. *In-situ* studies with electronic monitoring of feeding behaviour. *Pesticide Science* 49, 130–140.

Hoy, C., Vaughn, T. and East, D. (2000) Increasing the effectiveness of spring trap crops for *Leptinotarsa decemlineata*. *Entomologia Experimentalis et Applicata* 96, 193–204.

Hunt, T.E., Higley, L.G., Witkowski, J.F., Young, L.J. and Hellmich, R.L. (2001) Dispersal of adult European corn borer (Lepidoptera: Crambidae) within and proximal to irrigated and non-irrigated corn. *Journal of Economic Entomology* 94, 1369–1377.

Hurley, T.M., Babcock, B.A. and Hellmich, R.L. (2001) *Bt* corn and insect resistance: an economic assessment of refuges. *Journal of Agricultural Research and Economics* 26, 176–194.

IRAC (2005) Neonicotinoid Subcommittee Recommendations. http://www.irac-online.org/documents/projects/cni%20irac%20guidelines.pdf (accessed July 2005).

Jackson, R.E., Bradley, J.R. Jr and Van Duyn, J.W. (2004) Temporal and spatial production of bollworm from various host crops in North Carolina: implications for *Bt* resistance management. In: *Proceedings of the Beltwide Cotton Conference, San Antonio, Texas, 5–9 January 2004*. National Cotton Council of America, Memphis, Tennessee, pp. 1637–1640.

Jackson, R.E., Gould, F., Bradley, J.R. Jr and Van Duyn, J.W. (2006) Genetic variation for resistance to *Bacillus thuringiensis* toxins in *Helicoverpa zea* (Lepidoptera: Noctuidae) in eastern North Carolina. *Journal of Economic Entomology* 99, 1790–1797.

Jaffe, G. (2003) Planting Trouble: Are Farmers Squandering *Bt* Corn Technology? http://cspinet.org/new/pdf/bt_corn_report.pdf (accessed November 2007).

Leonard, M.D., Walker, H.G. and Enari, L. (1970) Host plants of *Myzus persicae* at the Los Angeles State and County Arboretum, Arcadia, California. *Proceedings of the Entomological Society of Washington* 72, 294–312.

Liu, Y.B., Tabashnik, B.E., Dennehy, T.J., Patin, A.L. and Bartlett, A.C. (1999) Development time and resistance to *Bt* crops. *Nature* 400, 519.

Liu, Y.B., Tabashnik, B.E., Dennehy, T.J., Patin, A.L., Sims, M.A., Meyer, S.K. and Carrière, Y. (2001a) Effects of *Bt* cotton and Cry1Ac toxin on survival and development of pink bollworm (Lepidoptera: Gelechiidae). *Journal of Economic Entomology* 94, 1237–1242.

Liu, Y.B., Tabashnik, B.E., Meyer, S.K., Carrière, Y. and Bartlett, A.C. (2001b) Genetics of pink bollworm resistance to *Bacillus thuringiensis* toxin Cry1Ac. *Journal of Economic Entomology* 94, 248–252.

Livingston, M.J., Carlson, G.A. and Fackler, P.L. (2004) Managing resistance evolution in two pests to two toxins with refugia. *American Journal of Agricultural Economics* 86, 1–13.

Luttrell, R.G., Wan, L. and Knighten, K. (1999) Variation in susceptibility of noctuid (Lepidoptera) larvae attacking cotton and soybean to purified endotoxin proteins and commercial formulations of *Bacillus thuringiensis*. *Journal of Economic Entomology* 92, 21–32.

Macarthur Agribusiness Report (2004) Evaluation of the Australian Cotton Industry Best Management Practices Program. http://www.cottonaustralia.com.au/library/research/Evaluation%20of%20the%20AustralianCottonIndustryBMP%20Program3.pdf (accessed December 2007).

Mahon, R.A., Olsen, K., Young, S., Garcia, K. and Lawrence, L. (2004) Resistance to *Bt* toxins in *Helicoverpa armigera*. *The Australian Cottongrower* (Dec. 2003–Jan. 2004), 8–10.

Marra, M.C., Pardey, P.G. and Alston, J.M. (2002) The payoffs to transgenic field crops: an assessment of the evidence. *AgBioForum* 5, 43–50.

MIASS (1994) *Agricultural Statistics*. Michigan Agricultural Statistics Service, Michigan Department of Agriculture.

MIASS (1995/1996) *Agricultural Statistics*. Michigan Agricultural Statistics Service, Michigan Department of Agriculture.

Michigan Potato Industry Commission (1998) *1998 Michigan Potato Pest Survey*. Michigan Potato Industry Commission, DeWitt, Michigan.

Misener, G., Bioteau, G. and McMillan, L. (1993) A plastic-lining trenching device for the control of Colorado potato beetle: beetle excluder. *American Potato Journal* 70, 903–908.

Monsanto (2004) A Guide to the 2004/05 Resistance Management Plan. http://www.monsanto.com.au/content/cotton/bollgard_ii_cotton/rmp.pdf (accessed November 2007).

Morin, S., Biggs, R.W., Sisterson, M.S., Shriver, L., Ellers-Kirk, C., Higginson, D., Holley, D., Gahan, J.J., Heckel, D.G., Carrière, Y., Dennehy, T.J., Brown, J.K. and Tabashnik, B.E. (2003) Three cadherin alleles associated with resistance to *Bacillus thuringiensis* in pink bollworm. *Proceedings of the National Academy of Sciences USA* 100, 5004–5009.

Mowry, T.M. (2001) Green peach aphid (Homoptera: Aphididae) action thresholds for controlling the spread of potato leafroll virus in Idaho. *Journal of Economic Entomology* 94, 1332–1339.

Moyer, D., Derksen, D. and McLeod, M. (1992) Development of a propane flamer for Colorado potato beetle. *American Potato Journal* 69, 599–600.

NASS (2004) Agricultural Chemical Usage, 2003 Field Crops Summary. Ag Ch 1 (04) a. http://usda.mannlib.cornell.edu/reports/nassr/other/pcu-bb/agcs0504.pdf (accessed November 2007).

National Research Council (NRC) (1986) Pesticide Resistance: strategies and tactics for management. National Academy Press, Washington, DC.

Nauen, R. and Elbert, A. (2003) European monitoring of resistance to insecticides in *Myzus persicae* and *Aphis gossypii* (Hemiptera: Aphididae) with special reference to imidacloprid. *Bulletin of Entomological Research* 93, 47–54.

NPC (2005) A Grower Approach to Resistance Management. Colorado Potato Beetle and Green Peach Aphid in Potato. http://www.nationalpotatocouncil.org/NPC/p_documents/document_210506112327.pdf (accessed August 2005).

Onstad, D.W. and Gould, F. (1998) Modeling the dynamics of adaptation to transgenic maize by European corn borer (Lepidoptera: Pyralidae). *Journal of Economic Entomology* 91, 585–593.

Peck, S.L., Gould, F. and Ellner, S.P. (1999) Spread of resistance in spatially extended regions of transgenic cotton: implications for management of *Heliothis virescens* (Lepidoptera: Noctuidae). *Journal of Economic Entomology* 92, 1–16.

Plapp, F.W. Jr, Campanhola, C., Bagwell, R.D. and McCutcheon, B.F. (1990) Management of pyrethroid-resistant tobacco budworms on cotton in the United States. In: Roush, R.T. and Tabashnik, B.E. (eds) *Pesticide Resistance in Arthropods*. Chapman & Hall, New York, pp. 237–261.

Radcliffe, E.B. and Ragsdale, D.W. (2002) Aphid transmitted potato viruses: the importance of understanding vector biology. *American Journal of Potato Research* 79, 353–386.

Radcliffe, E.B., Flanders, K.F., Ragsdale, D.W. and Noetzel, D.M. (1991) Potato insects – pest management systems for potato insects. In: Pimentel, D. (ed.) *CRC Handbook of Pest Management in Agriculture*. CRC Press, Boca Raton, Florida, pp. 586–621.

Roush, R.T. (1996) Can we slow adaptation by pests to insect-resistant transgenic crops? In: Persley, G. (ed.) *Biotechnology and Integrated Pest Management*. CAB International, Wallingford, UK, pp. 242–263.

Roush, R.T. (1997) Managing resistance to transgenic crops. In: Carozzi, N. and Koziel, M. (eds) *Advances in Insect Control: the Role of Transgenic Plants*. Taylor & Francis, London, pp. 271–294.

Roush, R.T. (1998) Two toxin strategies for management of insecticidal transgenic crops: can pyramiding succeed where pesticide mixtures have not? *Philosophical Transactions of the Royal Society of London. Series B, Biological Sciences* 353, 1777–1786.

Roush, R.T., Fitt, G.P., Forrester, N.W. and Daly, J.C. (1998) Resistance management for insecticidal transgenic crops: theory and practice. In: Zalucki, M.P., Drew, R.A.I. and White, G.G. (eds) *Pest Management – Future Challenges*. University of Queensland Press, Brisbane, Australia, pp. 247–257.

Sawicki, R.M. and Denholm, I. (1987) Management of resistance to pesticides in cotton pests. *Tropical Pest Management* 33, 262–272.

Shields, E.J., Hygnstrom, J.R., Curwen, D., Stevenson, W.R., Wyman, J.A. and Binning, L.K. (1984) Pest management for potatoes in Wisconsin – a pilot program. *American Potato Journal* 61, 508–516.

Showers, W.B., Hellmich, R., Derrick-Robinson, M. and Hendrix W. III (2001) Aggregation and dispersal behavior of marked and released European corn borer (Lepidoptera: Crambidae) adults. *Environmental Entomology* 30, 700–710.

Summerford, D.V., Hardee, D.D., Adams, L.C. and Solomon, W.L. (1999) Status of monitoring for tolerance to Cry1Ac in populations of *Helicoverpa zea* and *Heliothis virescens*: three-year summary. In: *Proceedings of the Beltwide Cotton Conference, Orlando, Florida, 3–7 January*. National Cotton Council of America, Memphis, Tennessee, pp. 936–939.

Suranyi, R., Longtine, C., Ragsdale, D. and Radcliffe, E. (1999) Controlling leafhoppers with below-label rates. *Valley Potato Grower* 64, 11–13 and 16.

Tabashnik, B.E., Patin, A.L., Dennehy, T.J., Liu, Y.B., Carrière, Y., Sims, M.A. and Antilla, L. (2000) Frequency of resistance to *Bacillus thuringiensis* in field populations of pink bollworm. *Proceedings of the National Academy of Sciences USA* 97, 12980–12984.

Tabashnik, B.E., Liu, Y.B., Dennehy, T.J., Sims, M.A., Sisterson, M.S., Biggs, R.W. and Carrière, Y. (2002a) Inheritance of resistance to *Bt* toxin Cry1Ac in a field-derived strain of pink bollworm (Lepidoptera: Gelechiidae). *Journal of Economic Entomology* 95, 1018–1026.

Tabashnik, B.E., Dennehy, T.J., Sims, M.A., Larkin, K., Head, G.P., Moar, W.J. and Carrière, Y. (2002b) Control of resistant pink bollworm by transgenic cotton with *Bacillus thuringiensis* toxin Cry2Ab. *Applied Environmental Microbiology* 68, 3790–3794.

Tabashnik, B.E., Carrière, Y., Dennehy, T.J., Morin, S., Sisterson, M.S., Roush, R.T., Shelton, A.M. and Zhao, J.-Z. (2003) Insect resistance to transgenic *Bt* crops: lessons from the laboratory and field. *Journal of Economic Entomology* 96, 1031–1038.

USEPA (1998) *The Environmental Protection Agency's White Paper on Bt Plant-pesticide Resistance Management*. No. 739-S-98-001. US Environmental Protection Agency, Washington, DC.

USEPA (2001) Biopesticides Registration Action Document – *Bacillus thuringiensis* Plant-Incorporated Protectants (10/16/2001). http://www.epa.gov/pesticides/biopesticides/pips/bt_brad.htm (accessed November 2004).

USEPA (2005) *Bacillus thuringiensis* Cry3Bb1 Protein and the Genetic Material Necessary for its Production (Vector ZMIR13L) in Event MON 863 Corn & *Bacillus thuringiensis* Cry1Ab Delta-Endotoxin and the Genetic Material Necessary for its Production in Corn (006430, 006484) Fact Sheet. http://www.epa.gov/oppbppd1/biopesticides/ingredients/factsheets/factsheet_006430-006484.htm (accessed November 2005).

USEPA-SAP (1998) *Subpanel on Bacillus thuringiensis (Bt) Plant-Pesticides, February 9–10, 1998. Transmittal of the final report of the FIFRA Scientific Advisory Panel on Bacillus thuringiensis (Bt) Plant-Pesticides and Resistance Management, Meeting held on February 9–10, 1998*. Docket Number: OPPTS-00231. US Environmental Protection Agency–Scientific Advisory Panel, Washington, DC.

Waller, B., Hoy, C., Henderson, J., Stinner, B. and Welty, C. (1998) Matching innovations with potential users, a case study of potato IPM practices. *Agriculture, Ecosystems and Environment* 70, 203–215.

Weisz, R., Smilowitz, Z. and Christ, B. (1994) Distance, rotation, and border crops affect Colorado potato beetle (Coleoptera: Chrysomelidae) colonization and population density and early blight (*Alternaria solani*) severity in rotated potato fields. *Journal of Economic Entomology* 87, 723–729.

Weisz, R., Smilowitz, Z. and Fleischer, S. (1996) Evaluating risk of Colorado potato beetle (Coleoptera: Chrysomelidae) infestation as a function of migratory distance. *Journal of Economic Entomology* 89, 435–441.

Whalon *et al.*, 2005. http://www.pesticideresistance.org/DB/index.html

Whalon, M., Mota-Sanchez, D., Hollingworth, R.M. and Duynslager, L. (2007) Resistant Pest Management: Arthropod Database. http://www.pesticideresistance.org/ (accessed February 2007).

Wright, R.J. (1984) Evaluation of crop rotation for control of Colorado potato beetles (Coleoptera: Chrysomelidae) in commercial potato fields on Long Island. *Journal of Economic Entomology* 77, 1254–1259.

Zhao, J., Grafius, E. and Bishop, B. (2000) Inheritance and synergism of resistance to imidacloprid in the Colorado potato beetle (Coleoptera: Chrysomelidae). *Journal of Economic Entomology* 93, 1508–1514.

6 The Politics of Resistance Management: Working Towards Pesticide Resistance Management Globally

G.D. Thompson[1], S. Matten[2], I. Denholm[3], M.E. Whalon[4] and P. Leonard[5]

[1]Dow AgroSciences LLC, Indianapolis, Indiana, USA; [2]USEPA Headquarters, Washington, DC, USA; [3]Rothamsted Research, Harpenden, UK; [4]Department of Entomology, Michigan State University, East Lansing, Michigan, USA; [5]Regulatory and Government Affairs, BASF Belgium, Brussels, Belgium

6.1 Introduction

Global and even national resistance management (RM) policies are rare or non-existent in most countries. The Food and Agriculture Organization of the United Nations (FAO), the World Health Organization (WHO), national and local officials, and manufacturers have not ignored the issue, but have generally reacted to specific cases of resistance after they have developed. There are several underlying factors which may help explain a lack of public policy for protecting key pest management tools critical to ensuring human and animal health, as well as our food supply. The first major factor is the relatively short time frame over which resistance to pesticides has been a problem. Modern pesticides and related resistance issues developed in the latter years of the 20th century, and plant-incorporated pesticides (PIPs), have been in commercial use only since 1995.

Another major factor is the complexity of the issues and circumstances surrounding resistance development. Pesticide resistance management (henceforth referred to as RM) is a blend of several basic sciences including, but not limited to, entomology, genetics, epidemiology, toxicology, ecology and mathematics (Denholm and Rowland, 1992; Thompson and Head, 2001). RM also involves societal investments in terms of institutions and social interactions including politics and divergent debate from various 'vested' positions (Dover and Croft, 1986; National Research Council, 1986). The majority of scientific theories can be tested, but with RM it is difficult to design definitive tests of resistance management strategies given the large spatial and temporal scales needed to observe the relevant selective responses (Denholm and Rowland, 1992). The large number of stakeholders and the complex and dynamic nature of resistance issues have produced excellent substance for debate, but have complicated consensus and policy development efforts. The diversity of the biological properties of the pests, cropping systems, pesticides and individual growers or pest control operators is extensive and, generally, more challenging than any other pest management aspect. The dynamic nature of resistance makes it a 'moving target' for policy makers and agrochemical producers. Lastly, pest resistance to date has been fairly limited and localized as new solutions for pest control (novel pesticides and integrated pest management (IPM) practices) have continued to come forward, allowing policy makers to focus on other critical agriculture and human health issues. These obstacles to policy development for RM will probably not disappear, but we believe that our working knowledge of resistance, and the experience gained with combating it, have advanced enough that RM should be an integral component of new policies relating to crop protection, and human and animal health.

Although there are a few notable examples of developing generic RM policies, most attempts have been recommendations to use a set of tactics or tools that address a specific problem. These recommendations are not true policies that articulate the issues and set direction and limitations. Clearly, what is needed for sound RM policy development is a long-term holistic and strategic view of the issues and solutions. Eventually, tactics need to be implemented at the local level; field by field or management unit by management unit. To facilitate and accelerate these changes, an overarching national, or even global, policy is needed that embraces an understanding of the huge value of pest management products to society and a commitment to sustain their availability and effectiveness.

This chapter reviews current and historic attempts at RM policy. Our goal is to articulate these policies in such a way that we can learn from failures and successes, and set the stage for future policy development. We outline the different stakeholders, their current activities and issues, and attempt to identify the unmet needs for successful RM. As a consequence of the authors' interests, there is a bias towards insects and insecticides; however, from a policy standpoint, most issues are generic and apply equally to other pesticide classes, including fungicides and herbicides. We conclude with some recommendations for future policy development, with the hope that the many resources deployed to combat resistance development in a resource-poor world can be put forward in a concerted cost-effective fashion to confront the challenges of RM.

6.2 Historical Review of Pesticide Resistance Management Policies

North America

In the USA, as in most areas of the world, there have been discussions on how to tackle resistance issues dating back to the early days of the 20th century, when pesticides such as calcium arsenate were widely used. IPM programmes began taking root in the 1950s and 1960s, in large part to attempt to solve resistance problems and to address over-reliance on crop protection products. In the 1980s, when serious problems of pyrethroid resistance were developing, the crop protection industry formed a Pyrethroid Efficacy Group, an early subgroup of the Insecticide Resistance Action Committee (IRAC, see below), to work with university and extension personnel, and develop regional response programmes that had some degree of success (Jackson, 1986). A national debate in the mid-1980s generally concluded that regional programmes were the most effective, and that national programmes or involvement by the Environmental Protection Agency (EPA) would be less effective (Hawkins, 1986; Johnson, 1986). However, for most pests, there have been very few concerted efforts directed at RM and even fewer that were proactive.

Conventional insecticides

Following the adoption of the Food Quality Protection Act (FQPA) in 1996, pesticide resistance management assumed a higher national profile in the USA. This was because the FQPA led to the elimination of many uses of two broad-spectrum pesticide groups, organophosphates and carbamates, which could not meet newer safety standards or re-registration cost hurdles, but none the less were key components of RM programmes. Loss of these broad-spectrum pesticides inevitably put more selection pressure on pests to develop resistance to the remaining pesticide groups, especially the narrower-spectrum and more environmentally benign ones, and transgenic crops. Without broad-spectrum pesticides such as organophosphates, minor pests can become major problems and resistance risks. Pesticide resistance is, therefore, likely to increase in importance as the diversity of active ingredients decreases and the target spectrum of remaining products is narrowed.

In contrast to the mandatory RM strategies required for *Bt* plant-incorporated pesticides (PIPs) (see below), both the EPA and the Canadian Pest Management Regulatory Agency (PMRA) endorse and promote use of voluntary resistance management strategies for all other pesticides (conventional, biochemical and microbial pesticides) to mitigate risks of resistance. Beginning in Canada in October 1999 and the USA in June 2001, PMRA and EPA, respectively, developed voluntary RM labelling guidelines based on the rotation of pesticides with different modes of action as part of a joint activity under the North American Free Trade Agreement (NAFTA), in conjunction with the Risk Reduction Subcommittee of the NAFTA Technical Working Group on Pesticides. Both countries believed that a harmonized approach based on such rotations would help slow down or arrest the development of resistance. These guidelines were published as EPA Pesticide Registration Notice 2001-5 (USEPA, 2001b) and Canada Regulatory Directive DIR 99-06 (Health Canada, 1999). They provide a numerical system of classification to identify the mode of action of the pesticide on the front panel of the label and to provide resistance management labelling statements in the 'use directions'.

The numerical mode of action (MOA) classification systems were developed in consultation with three international industry technical groups whose focus is pest resistance: the IRAC, the Herbicide Resistance Action Committee (HRAC), and the Fungicide Resistance Action Committee (FRAC). Other organizations, such as the Weed Science Society of America (WSSA), made important contributions, and the fungicide mode of action classification was originally modelled on a system already developed in Australia. MOA classifications are updated on an annual basis and are posted on each organization's web site (addresses shown later). The labelling statements encourage users to: (i) rotate between pesticides with different modes of action; (ii) monitor for loss of field performance; (iii) use IPM programmes that include resistance management; (iv) use up-to-date technical information regarding resistance management, provided by technical representatives from industry, academia, extension and consultants; and (v) report any suspected instances of resistance. The goal of this voluntary policy is to educate users away from control regimes based wholly or largely on the successive use of only one chemical group, a strategy greatly increasing the likelihood of resistance. This approach to resistance management is technically sound, practical, feasible and benefits pesticide manufacturers, pesticide users and the public. EPA and PMRA have stated a desire to work with the pesticide industry to embrace this approach and to implement the Pesticide Registration Notice 2001-5 and Regulatory Directive 99-06 for all relevant products. Both the USA and Canada believe this approach is an important element of international harmonization that could alleviate resistance problems and support a unified approach to joint registration decisions in any, or all, NAFTA countries and even worldwide.

Implementation of these voluntary guidelines by the agrochemical industry has been more limited in the USA than in Canada (Matten and Beauchamp, 2004). However, efforts are under way by industry, academia, and producer organizations to use the numerical classification system and resistance management labelling recommendations, and to develop educational materials that stress the importance of resistance management. Many state pesticide education and IPM coordinators use the MOA classification systems in their educational materials, for example in Florida and New York. The National Potato Council has recently issued guidance to its growers on neonicotinoid resistance management that includes the use of the insecticide MOA classifications and the recommendation of rotation of mode of action to reduce the likelihood of resistance. IRAC has also issued neonicotinoid resistance management guidance and maintains the MOA classification listing.

In addition to the voluntary resistance management labelling programme, the EPA has other regulatory activities that help address pest resistance problems. Under current policy, EPA can only authorize emergency exemptions for the purposes of resistance management when resistance is documented to the registered alternative(s) and continued use of the latter will result in significant economic loss. Approximately 400–600 emergency exemptions are approved each year by EPA, with 30 to 50% citing resistance problems as the rationale for the emergency condition. Various stakeholders have been interested in EPA amending its guidance to allow for emergency exemptions to be issued for the purposes of resistance management. Under current procedures, if there is at least one available registered alternative that is effective enough to prevent significant economic losses, then the situation is generally not considered an emergency regardless of whether or not the alternative is considered to be vulnerable to the development of resistance by the target pest. Waiting until there are no alternatives has been criticized for the potential to exacerbate resistance situations. However, on a practical level, the EPA has issued emergency exemptions for multiple chemicals where resistance management was one reason for the request. One example of effective resistance management under current EPA policy was the simultaneous authorization in 1996 of pyriproxifen and buprofezin for controlling the whitefly, *Bemisia tabaci*, on cotton in Arizona following serious control failures with synergized pyrethroids (Dennehy and Williams, 1997; Dennehy and Denholm, 1998). Other examples were the authorization in 1999 of coumpahos for Varroa mite control in 40 states due to fluvalinate resistance, and the authorization of ApiLife VAR (thymol, methanol, eucalyptus) in 15–20 states during 2004 due to coumaphos resistance. Examples involving fungicides were the emergency authorization of three products to combat late blight on potato, and exemptions in 1998 for use of tebuconazole, myclobutanil and fenarimol to control powdery mildew on hops. These examples are in contrast to the general EPA policy not to allow emergency exemptions for the purposes of resistance management even if there is only one effective product. This policy has concerned many stakeholders who believe that it is short-sighted and will put additional pressure on use patterns with few products and particularly on specialty or minor crop industries, which tend to have fewer registered alternatives and are more prone to resistance problems.

Bt *plant-incorporated pesticides*
During the 1990s, the introduction of transgenic plants engineered to produce specific insecticidal proteins isolated from the ubiquitous soil bacterium, *Bacillus thuringiensis* (*Bt*), focused the debate on RM to a level never seen before in the USA. This activity was triggered by the submission of an experimental use permit application by Monsanto Company for its cotton engineered to produce the Cry1Ac protein (Bollgard® cotton) to control certain species of lepidopteran pests. Several public interest groups expressed a concern that commercialization of *Bt* cotton would increase the likelihood of *Bt* resistance in pests such as *Heliothis virescens* (tobacco budworm), *Helicoverpa zea* (corn earworm or cotton bollworm) and *Pectinophora gossypiella* (pink bollworm) because of high-level and season-long constitutive expression of the *Bt* protein(s). There were also concerns over the impact on organic agriculture that relies, in many instances, on *Bt* microbial sprays. These issues were repeatedly and emphatically brought to the EPA's attention by public interest groups. In 1998, EPA stated that development of resistant insects would constitute an adverse environmental effect and made the determination that protecting the susceptibility of *Bt* was in the 'public good' (USEPA, 1998, 2001a). The loss of *Bt* as an effective pest management tool in maize, cotton or other crops would probably have negative consequences for the environment to the extent that growers would probably shift to the use of higher-risk pesticides, and a valuable tool for organic farmers would be affected. Sound RM tactics were needed to prolong

the life of *Bt* insecticides and provide economic and environmental benefits to growers, consumers and industry. In 1998, EPA stated that its RM policy for *Bt* PIPs was to mitigate the risks of resistance to *Bt* and sustain the benefits of this technology (USEPA, 1998), and, in 2001, it published an extensive reassessment of the risks and benefits of *Bt* PIPs (USEPA, 2001a).

IRM for *Bt* PIPs has focused on the use of the high-dose/structured refuge strategy to combat resistance to specific proteins (Roush, 1997a,b; Gould, 1998a,b). A high dose has been defined as 25 times the protein concentration needed to kill susceptible insects (Scientific Advisory Panel, 1998, 2001). A high dose and the planting of a structured refuge (specific plantings of a non-*Bt* crop intended to produce susceptible insects that can randomly mate with resistant insects) is intended to kill resistance heterozygotes and dilute the frequency of resistance alleles in the population.

EPA demands that the *Bt* PIP registrants require growers to plant specific non-*Bt* refuges for all but some dual *Bt*s and have an annual resistance monitoring programme, a remedial action programme and a compliance assurance programme. Results of the monitoring and compliance assurance programmes should be reported to EPA annually (USEPA, 2001b). These RM strategies are designed to be sustainable and require modification as our understanding of the science improves. The mandatory RM programmes for *Bt* PIPs are both proactive and unprecedented in detail. RM requirements for *Bt* maize and *Bt* cotton PIP products are described in EPA's *Bt* PIPs reassessment document (USEPA, 2001a) and in numerous publications that discuss various aspects of the development of RM strategies, resistance monitoring programmes, stakeholder involvement and long-term sustainability of RM strategies for *Bt* PIPs (e.g. Glaser and Matten, 2003; Matten and Reynolds, 2003; Matten *et al.*, 2004). One might conclude that the current lack of insect resistance to *Bt* PIPs is partly due to the specific institution of mandated RM by EPA, and the support and adoption by industry and the growers of these specific RM requirements (Tabashnik *et al.*, 2003).

In conclusion, there have been a number of efforts in North America to address pest resistance problems. Some of them have been voluntary, and some of them have been mandatory. Efforts to combat resistance proactively to *Bt* proteins expressed in *Bt* maize and cotton have been highly focused and very successful. Despite the great focus on managing resistance to *Bt* crops, less attention is still being paid to managing resistance to conventional pesticides. Greater focus on efforts to promote a holistic approach to managing resistance across technologies is advisable. As yet, however, no succinct policies are in place to encourage adoption of such approaches.

Europe

Policy development relating to pesticide use and resistance management in Europe is increasingly becoming centralized by attempts to harmonize procedures for approval and regulation across countries within the EU. The EU registration process, driven by Commission Directive 91/414, recognizes the importance of resistance and requires registrants to address the risk of resistance development as part of dossiers submitted for EU registration (European Commission, 1993; MacDonald *et al*., 2003). This is the first attempt to formally address resistance management at the regulatory level in the EU, and has developed in conjunction with the European and Mediterranean Plant Protection Organization (EPPO), which works with regulatory authorities and industry to develop guidance on resistance risk analysis and RM strategies. A guideline (PP 1/213(2)), first published by EPPO in 2000 and revised in 2003 (Anon., 2003), is now increasingly used by both applicants and regulatory authorities when evaluating resistance risk for registration purposes. In addition, EPPO has established a Resistance Panel on Plant Protection Products to assist with the implementation of PP 1/213(2) and to extend expert advice to member countries (EPPO, 2007a).

There is considerable concern in the EU over what effect the re-registration of older active ingredients, under Directive 91/414 EEC, will have on product diversity and sustainable use of plant protection products (European Commission, 2001, 2004). Loss of products arising both from the review of existing products under Directive 91/414 EEC linked with loss through lack of commercial support due to regulatory costs when additional safety data are required to support uses, especially minor uses, means that effective resistance management becomes more challenging. It is important therefore when interpreting guidelines that pragmatism is applied to ensure that further loss of products and uses does not occur.

Product use restrictions at the national level can potentially constitute a barrier to efficient implementation of RM strategies. For economic reasons, registrants are less likely to invest in dossiers to authorize use on minor or specialty crops at the national level, where returns on investments will be most limited. As a result, these registered use patterns are likely to be unsupported and therefore lost. The problem was officially recognized, and use derogations were granted for a number of products where no alternatives would remain (European Commission, 2003). However, these exemptions expire in 2008, and, at the time of publication, no means of prolongation is foreseen. The fundamental issue therefore remains that the resistance management toolbox is rapidly being depleted. To put this in context, of the 952 registered compounds which were eligible for re-registration in 1995, 589 have already been de-registered, and only 69 new compounds have been approved.

New active ingredients, modes of action, and use patterns are therefore badly needed to reverse this trend. However, the current rate of approval for new active ingredients in the EU (approximately five per year) is not likely to solve the problem in the foreseeable future. Resistance management in the EU will therefore be both increasingly important and difficult as product diversity is drastically reduced. An appreciation of the impact of re-registration on resistance risk should therefore be combined with pragmatism by national regulatory authorities, to avoid situations where further loss of product and registered use diversity prevents establishment of meaningful resistance management strategies. The slow adoption rate of genetically modified crops, particularly those with PIPs (*Bt*), is further limiting pest control options in the EU.

This is not limited to minor crops. There are already cases in the EU where viable resistance management alternatives are not available for major crops. An emerging resistance problem with pollen beetles (*Meligethes* spp.) on oilseed rape in northern Europe provides a timely reminder of the potential for resistance problems in major EU crops. With pyrethroid resistance now widespread in northern Europe, and very few non-pyrethroids registered for pollen beetle control, management options are simply running out. The conclusions of a recent international workshop on this problem can be found on the EPPO web site (EPPO, 2007b).

Compliance with guideline PP 1/213(2) can prove problematic for other reasons. Relatively few countries within the EU possess the technical expertise in resistance research needed to interpret registration dossiers and to judge whether risk assessment analyses and mitigation tactics (when appropriate) are adequate. The EPPO Resistance Panel provides an important source of technical advice. As in North America, guidelines produced by the international Resistance Action Committees (including MOA classification schemes) offer considerable help. However, some agencies openly question whether organizations representing only the agrochemical industry are sufficiently independent in this respect. The UK has taken an important initiative by forming national Resistance Action Groups addressing insecticides (IRAG-UK), fungicides (FRAG-UK), herbicides (WRAG) and rodenticides (RRAG). These groups have a wide membership including researchers, regulators, advisors, grower representatives, non-governmental organizations and agrochemical companies, and are able to focus on issues of greatest concern within the UK (MacDonald *et al.*, 2003). Similar approaches are now being adopted by a number of other EU member countries.

Australia

This country provides some of the most enlightened examples of the progressive development and adoption of RM policies and tactics. These developments have largely been driven by the need to combat insecticide resistance in cotton pests. A climate favourable to insects and highly productive cropping systems that are concentrated in irrigated valleys is a good recipe for developing resistance, which has resulted in resistance to several chemical groups by the bollworm, *Helicoverpa armigera*. These problems commenced with the appearance of resistance to DDT in the early 1970s, which contributed to the collapse of cotton cultivation and the economic decline of the Ord Valley region of Western Australia (Wilson, 1974). Subsequent problems with resistance to pyrethroids led to the inception of an acclaimed RM strategy that limited use of certain insecticide groups to 'windows' during the cotton season and recommended alternation between groups (Sawicki and Denholm, 1987; Forrester *et al.*, 1993). This strategy has continued to the present day with considerable revisions to accommodate new chemicals becoming available, and to tailor recommendations to regions differing in pest incidence and crop phenology. These days the strategy is overseen by an official Transgenic and Insect Management Strategy (TIMS) Committee, composed of government researchers from federal- (e.g. CSIRO) and state-run (e.g. New South Wales Department of Agriculture) organizations, industry technical experts, crop consultants and growers, which develops and publishes regionally adapted recommendations (Farrell, 2006). Considerable financial and logistical support is provided by the Cotton Research and Development Corporation of Australia. Australians are generally accredited with developing the 'window' approach to resistance management, although others have promoted rotations in the literature (Plapp *et al.*, 1979; Roush, 1989), with strict time periods when certain modes of action can be utilized. The programme remains voluntary, but there is near 100% compliance since there is a well-developed consultant network for a relatively few large growers that understand the importance of the issue. In addition, the region is organized into water catchment districts that further facilitate timely communication, organization, and enforcement of agreed tactics. This *de facto* volunteer policy evolved over time, but is working well for the Australians and is an outstanding example of communication and buy-in at the grower level that has resulted in successful implementation.

As its name suggests, the TIMS Committee also oversees RM policy and recommendations for *Bt* transgenic cotton, which was introduced in Australia at approximately the same time as in the USA. The policy follows similar principles relating to high levels of expression and the planting of refuge crops capable of producing sufficient *Bt*-susceptible *H. armigera* to dominate the mating with any survivors of *Bt* crops and thus maintain resistance at low levels. As in the USA, compliance with the RM measures is mandatory, although there is flexibility in the design and size of refuges depending on the type of refuge crop and how it is managed (Fitt, 2003).

Other countries

A complete review of successes and failures with developing RM policies elsewhere in the world is beyond the scope of this chapter. However, some selected examples of coordinated ventures involving governments, research organizations, industrialists, and growers serve to highlight general principles and challenges to be faced in the future. For insecticides, the evolution of policies for RM in many parts of the world has tended to be dominated by problems with cotton production, a crop especially vulnerable to pest damage and often very reliant on the use of pesticides. A review of factors promoting or constraining structured approaches to RM in cotton is provided by Sawicki and Denholm (1987). The challenges are exemplified well by events in Africa, which has a very large and diverse number of agricultural systems with cases of success and failure regarding resistance development in pests. Historically, a few countries, such as Egypt, utilized a government purchase and distribution system for insecticides, which is the opposite of the practice of 'every grower for himself' that operates in most parts of the world. The centralized system in Egypt led, in the aftermath of serious resistance problems with leafworms on cotton in the 1970s, to a tightly controlled strategy for the release of synthetic pyrethroids dictated at national level by the Egyptian Crop Protection Committee (Sawicki and Denholm, 1987). Similar experience of control failures against cotton pests in Zimbabwe promoted voluntary compliance with a nationwide strategy for alternating pesticides over space and time, and restricting the use of new products appearing on the market (Blair, 1986). A contrasting and pragmatic approach was conceived for managing cotton pests in French-speaking West Africa by local government agencies working in collaboration with the Centre de Cooperation Internationale en Recherche Agronomique pour le Dévellopement (CIRAD) in France. This entailed calendar sprays of insecticide mixtures tailored to control the cotton pest complex as a whole with a single application. The strategy was sufficiently versatile to be implemented successfully across 11 African countries with different climates, ecology and pests (Cauquil, 1985). Although not specifically designed with RM in mind, this programme demonstrated the feasibility of adopting and disseminating standardized policies for pesticide use over large areas in developing countries, and consequently the opportunities to incorporate more specific RM recommendations.

Coordinated programmes such as those introduced in Egypt, Zimbabwe and West Africa have, in theory at least, several advantages including uniform training and information distribution, and potential for area-wide application of tactics. Unfortunately, some of the worst resistance situations have also resulted from government-led programmes in which purchasing agents, policies, funding, and crop advisor recommendations were not coordinated with good RM practice. On occasions, farmers have been provided with only one type of control agent, creating intense selection pressure for resistance. Purchasing agents were at times mandated by law or tempted by graft to purchase the least expensive product regardless of its effectiveness. Customizing programmes for different geographies and crops has also turned out to be logistically impossible at the national level in many situations (Griffiths, 1984). One outstanding exception has been the evolution, from humble beginnings, of an integrated nationwide RM programme for insect pests on cotton in India. Following a succession of resistance problems beginning in the 1970s that had proved economically devastating for subsistence farmers, work during the 1990s laid the foundations of a system for rationalizing pesticide use that began at village level and by 2001 extended to four Indian states. Subsequent funding from the Indian government, supplemented by support from overseas aid agencies in the UK, has since led to a rapid expansion of the RM programme to virtually all cotton-producing regions of India (Russell et al., 2003). Statistics show the average number of insecticide applications to have reduced by at least 50%, while yields have increased and the profitability of cotton production has improved by up to 85% during the same time. This dramatic reversal in fortunes has been achieved despite formidable logistical constraints and has had a profound influence on organizations that determine agricultural policy in India. It also serves as a template for embedding RM into reformed policies and practices elsewhere in the developing world.

Global initiatives

In principle, organizations such as FAO and WHO are uniquely placed to take an international lead in addressing resistance development and in coordinating extension and training programmes (National Research Council, 1986). They have provided good general recommendations for sustainable programmes and resistance avoidance, and, while a specific policy has not been articulated, the tactics that such a likely policy would endorse have been stated. WHO has invested significant resources in monitoring resistance in malaria vectors and has voiced concern over the critical shortage of management tools and modes of action that are available for pesticide rotation strategies. There was also a short-lived group called the International Organization for Pesticide Resistance Management (IOPRM) in the early 1990s, led from both the USEPA and the agrochemical industry, that attempted to provide some overarching coordination and policy development, but which could not achieve sufficient momentum to influence new funding, policy, and cooperation at the time. From all of these experiences, it is still not clear what organization has what role, and, if there were an international policy, how it would be implemented in the face of national vested interests and divergent approaches around the world for approving and regulating pesticides, as well as distributing them to end-users.

6.3 Current Policy Advocates and Stakeholders

The crop protection industry

Agrochemical companies have much to gain or lose from pest resistance, since it is the sustained sales of their products that are at stake. All companies should be aware of resistance issues, and some multinational ones have dedicated in-house programmes to monitor resistance to their products and to develop product-specific RM recommendations. Following some initial distrust over their motives, the Resistance Action Committees (RACs) have steadily become recognized as authoritative sources of RM information and tools. The overall mission of the RACs is to facilitate communication and education on pesticide resistance and to promote the development of RM strategies in crop protection and disease control to maintain efficacy, agricultural sustainability, and improved public health. In terms of organizational structure, the RACs are task forces, or working groups, of CropLife International, and, as such, are recognized by FAO and WHO as advisory bodies.

Further insights into the structure and activities of the RACs can be provided by taking IRAC as a specific example. Overall coordination is through an international committee (IRAC International), but there are also several national or regional committees that take a more parochial view of resistance issues, with information disseminated through meetings, workshops, educational materials, and web sites. Some of these national committees have proved outstandingly successful (e.g. IRAC-Brazil), while others (e.g. IRAC-Pakistan) have proved more transient due to local organizational difficulties. They are comprised primarily of key technical personnel from the agrochemical companies affiliated with CropLife through membership in the relevant national associations (European Crop Protection Association, CropLife America, etc.). In the USA, the Agricultural Biotechnology Steering Committee (ABSTC) was formed to develop consensus responses to technical questions on PIPs, and resistance management subcommittees of ABSTC address questions and research specific to PIPs in consultation with IRAC-US. The IRAC groups on occasion provide funding for resistance management projects around the world. These are generally driven or coordinated by the local country group, or, in some cases, a specific project group is set up to lead and ultimately place results and findings into the public domain. Examples of these have been the long-term evaluation of RM strategies for mosquitoes in Mexico, the monitoring of pyrethroid resistance in *H. armigera* in West African cotton and in *H. virescens* and *H. zea* in the USA, and the monitoring of resistance in the codling moth, *Cydia pomonella*, in Eastern Europe. Other groups have been established to assist with stewarding specific groups of pesticides. The Pyrethroid Efficacy Group was an early example, and, recently, a new project group was set up within IRAC International and IRAC-US to coordinate management of the expanding group of neonicotinoid insecticides. Other activities focus on education, communication, and regulatory approvals, as well as providing expert technical support.

Further information on the RACs and their outputs is available from their respective web sites:

- IRAC – www.irac-online.org/
- FRAC – www.frac.info/frac/index.htm
- HRAC – www.plantprotection.org/HRAC/

Public research organizations and funding agencies

Around the world, research on pesticide resistance is conducted by a diverse community of scientists based in universities, research institutes, national and state agricultural bodies such as the US Department of Agriculture (USDA), and extension agencies. Funding sources are equally diverse in terms of level of support and consistent commitment. Fundamental research on the evolutionary ecology and genetics of resistance, as well as work to disclose underlying mechanisms, is usually supported by research councils or science foundations that operate nationally, but provide some scope for international collaboration. Government departments (covering agriculture, health and overseas aid, and including regulatory agencies) generally take a more strategic view based on political expediency, while commodity groups, funded largely by growers, are understandably keen to promote immediate, practical solutions to problems associated with specific pests or crops. In addition, European researchers have access to substantial funds administered and distributed at the level of the EU, and there have been examples of these funds being accessed to coordinate research on resistance across member countries. One successful example was a concerted action termed ENMARIA that brought scientists from 13 EU countries together for 3 years to exchange knowledge and expertise on insecticide resistance monitoring, mechanisms, and management (Denholm and Jespersen, 1997). More recently, the EU has awarded funds to build a European-wide network of researchers that has provision for supporting collaborative work on resistance to insecticides, fungicides, and herbicides within the context of changing EU policies on crop protection (www.endure-network.eu). RM is also within the remit of several charitable organizations, including the Welcome Trust and the Gates Foundation, especially in relation to problems affecting human health.

Agrochemical companies have increasingly become involved in supporting research in the public sector, providing funds individually, collectively (e.g. through the RACs), or in collaboration with other agencies. Alongside the USDA and other organizations, IRAC has sponsored an ambitious initiative to generate an interactive, web-based database of known cases of arthropod resistance to insecticides and acaricides. This Arthropod Pesticide Resistance Database (APRD) was constructed by staff at Michigan State University (MSU) in the USA and follows up earlier work at the University of California to collate reports of resistance in the scientific literature (Georghiou and Lagunes-Tejeda, 1991). To date, the APRD has documented more than 7000 cases of arthropod resistance and offers automated case reporting and data retrieval on the World Wide Web (www.pesticideresistance.org). In collaboration with the WSSA, HRAC supports a similar database on herbicide resistance that currently holds records of 314 resistant biotypes representing 183 weed species (www.weedscience.org/in.asp).

In the USA and many other countries, efforts to implement resistance management have been promoted through cooperative efforts that bring researchers, crop consultants, commodity organizations and extension personnel together with pesticide industry and government personnel. In the wake of a US National Research Council Board on Agriculture initiative (National Research Council, 1986), several specific programmes were initiated. Initially, a Western Regional Resistance Management Coordinating Committee funded through USDA and the Cooperative State Research, Education, and Extension Service (CSREES) was initiated with representatives from all regions in the USA and delegates from Canada and Mexico. WRCC-60, together with MSU, has published the *Resistant Pest Management Newsletter* (whalonlab.msu.edu/rpm/index.html) bi-annually for the last 14 years. This newsletter is currently supported by MSU, USDA/CSREES, and IRAC International; and publishes from 40 to 140 articles, abstracts, and updates on resistance and RM development from around the world annually.

The breadth of potential funding sources for resistance work should, in principle, facilitate a continuum of research from fundamental to highly applied projects and a means of translating results of basic research as rapidly as possible into practical solutions. However, in many cases, the lack of any group or organization taking a 'top-down' strategic view of priorities and activities (especially from an international perspective) means that work undertaken is often piecemeal and fragmented. The main constraint in this area is probably not a shortage of resources per se (although this is frequently a challenge), but lack of sufficient coordination between funding agencies to avoid unnecessary duplication of effort and to enable research on RM to be conducted over more ambitious and realistic experimental scales.

Regulatory organizations

In the USA, there has been a shift of emphasis from the view that resistance management is too complex to regulate (Johnson, 1986) to a mix of mandatory requirements for PIPs and strong voluntary guidelines for conventional pesticides (Matten and Beauchamp, 2004). The pioneering approach of the EU was described previously in this chapter, and most other areas have relied on existing use pattern regulations or are just starting to develop guidelines. Global awareness of resistance has unquestionably been reinforced by the international focus placed on management of insect resistance to *Bt* proteins expressed in transgenic crops, especially *Bt* maize and *Bt* cotton products. Australia, the Philippines, India, Mexico, and South Africa all have mandatory RM strategies similar to those in the USA. China and India pose particular challenges for implementing a high-dose/refuge strategy, but numerous discussions have been held on how these challenges can be overcome (e.g. Kranthi *et al*., 2003). None of these countries, however, has mandatory RM programmes for pest control products other than *Bt* crops. Only the EU has so far adopted mandatory resistance monitoring and resistance risk analyses as part of the approval process for new and existing pesticides, and national agencies, such as the Pesticide Safety Directorate of the UK Department for the Environment, Food, and Rural Affairs, are drawing up guidelines for the implementation of this legislation (MacDonald *et al*., 2003).

Growers and commodity groups

Most people now consider that RM is a shared responsibility, but Benbrook (2004) has argued that the primary accountability lies with the grower, or their consultant, not with industry, universities, government agriculture departments, or regulators. It is certainly true that for RM programmes to be successful they have to be implemented at the individual field level. However, it is usually also true that RM tactics need to be implemented across regional populations of pests. This requires coordination beyond the individual grower, who is generally powerless to influence the dynamics of resistance selection on his or her own.

Fortunately, most countries have grower organizations that do exert a high level of RM coordination. Many of these have adopted a high profile in addressing resistance and often in providing funds for research with a direct bearing on the commodities they represent. Once again, cotton production provides a number of examples of grower engagement with groups, such as the National Cotton Council and Cotton Incorporated in the USA, having assumed an integral role in supporting researchers and advocating the deployment of RM strategies. The Australian Cotton Research and Development Corporation is one of the main funders of resistance research in that country and provides an infrastructure for coordinating the efforts of scientists, consultants, and extension personnel. In some European countries, the levy boards representing commodities such as cereals, potatoes, and horticulture enter into partnerships with researchers, industry and government departments to co-fund work that benefits greatly from the pluralist perspective that these consortia provide.

Another very important development has been the increased willingness of commodity groups to collaborate and address pest problems that transcend individual types of crops; for example, insects with wide host ranges that attack several crops simultaneously or cycle between successive crops as the season progresses. Also, grower organizations are often the most effective means of disseminating information on RM through their web sites and in-house trade publications and workshops.

6.4 Requirements and Responsibilities for Resistance Management Policy

Obtaining consensus over RM strategies and tactics appears to be a difficult assignment until one moves on to who should fund the activities. Historically, educational activities have been conducted by universities, public extension services, and companies introducing new products. However, as resources have become more constrained in all of these sectors, there has been an obvious trend to drop rather than add activities. Many have suggested that industry or the company registering the products should bear the bulk of the expense. However, private companies have to make a profit to stay in business, so costs are passed on to the growers. If the support requirements are not justified by the market, the products will not be developed or will need to be priced higher in countries with RM requirements, making growers in these countries less competitive on a global basis. The fundamental balance to be struck in terms of resources is between the roles of RM in maintaining profits (through sustained sales of important products) and in ensuring the sustainability of pest management (by preserving pest susceptibility). The former is clearly a company responsibility, but the latter is a much more collective one. Governmental or multi-national organizations, such as USDA, CSIRO, FAO and WHO, have numerous stewardship programmes aligned towards sustainability that incorporate RM considerations. Likewise, commodity organizations often fund sustainability efforts through commodity levies or surcharges. This topic is still controversial and needs considerably more debate. The authors believe that all stakeholders have a contribution to make, but that levels of the accountability and resource participation need greater clarification.

Even a cursory review of the scientific and trade literature around crop protection reveals that large amounts of private and public resources are being used to address RM issues. Findings from symposia and workshops on resistance management, such as the one conducted by the Council for Agricultural Sciences and Technologies (CAST, 2004), indicate that current resources to address resistance management in a proactive fashion are inadequate and that there is no unified policy that directs resources towards RM in the context of IPM and sustainable agricultural systems. Obtaining additional resources in the current socio-economic climate will be difficult at best, and trying to establish new agricultural policies that incorporate proactive approaches to RM will be challenging. Improved coordination and focusing of existing resources is the only realistic avenue to pursue. On simpler issues, such as education, there are often multiple groups attempting to meet the same needs and competing for the same resources without the coordination and efficiency to make these efforts cost-effective. If these groups, representing both the public and private sectors, could forge more productive working relationships, it is more likely that a mutually acceptable and uniform RM policy can be developed and implemented. However, each group's roles and responsibilities first need to be better defined, so that more holistic working relationships can be established, and the goals of RM can be achieved (Carter, F.L., 2004; Carter, R., 2004).

The CAST conference held in 2003 brought many stakeholders together, and participants concluded that RM is very important to the sustainability of agricultural production systems (CAST, 2004). Delegates at this meeting agreed that achieving proactive RM is a desirable goal, but achieving it is a complex process that requires extensive input and commitment by all. From the US perspective, it was agreed that both EPA and USDA play important roles in dictating and achieving RM objectives. In the EU, it took over a decade to develop policy that is transforming attitudes towards RM and is forcing a reconciliation of vested interests that have hampered a concerted approach to this subject in the past. One consequence of these developments is that it has become clear that while much is known about insect pests and the crops they damage, we are still woefully ill-equipped to predict the probability of resistance arising and the likely success of mitigation tactics. Maybe this requires a paradigm shift in approaching these problems since, in these complex biological systems, there will always be significant variability that is not easily reduced even with significant research efforts. The key question stated in pragmatic terms becomes, 'Should we wait until resistance occurs or knowledge is developed and react accordingly, or should we routinely institute more proactive practical resistance management practices that have inherent uncertainty as to their value and probability of success?' Given that pesticides and PIPs continue to underpin crop protection globally and that chemicals and traits surviving the current rigorous approval processes must, by necessity, pose low risk to human health and the environment, then RM is a logical step towards protecting the benefits they offer to mankind. Even the most intuitive economic considerations suggest that the costs of resistance should be incorporated into production agriculture, horticulture, stored product pest management or animal and human health protection, resulting in compelling arguments for approaching RM in a more intelligent, integrated and proactive manner.

Acknowledgements

The views expressed in this chapter are those of the individual authors and do not necessarily reflect the views and policies of Dow AgroSciences, Michigan State University, BASF, the US Environmental Protection Agency, the Institute of Horticultural Research, or Rothamsted Experiment Station. The use of trade names does not imply endorsement by the US Government.

We thank Alan McCaffery, Sheryl Reilly, David Richardson and Nick Storer for valuable discussions and comments on previous drafts of the manuscript.

References

Anon. (2003) Efficacy evaluation of plant protection products. PP1/123(2) Resistance Risk Analysis. *EPPO Bulletin* 33, 37–63.

Benbrook, C. (2004) Stakeholder roles in resistance management: time to get with the program. In: *Management of Pest Resistance: Strategies Using Crop Management, Biotechnology, and Pesticides*. CAST Special Publication No. 24. Council for Agricultural Sciences and Technologies, Ames, Iowa, pp. 128–132.

Blair, B.W. (1986) Strategies to minimize resistance in arthropod pests to acaricides and synthetic pyrethroid insecticides in Zimbabwe. In: *IVe Congrès sur la Protection de la Santé Humaine et des Cultures en Milieu Tropical*. Chambre de Commerce et d'Industrie, Marseilles, France, pp. 222–227.

Carter, F.L. (2004) Role of producers in management of resistance. In: *Management of Pest Resistance: Strategies Using Crop Management, Biotechnology, and Pesticides*. CAST Special Publication No. 24. Council for Agricultural Sciences and Technologies, Ames, Iowa, pp. 92–93.

Carter, R. (2004) Role of stakeholders in resistance management: crop consultants. In: *Management of Pest Resistance: Strategies Using Crop Management, Biotechnology, and Pesticides*. CAST Special Publication No. 24. Council for Agricultural Sciences and Technologies, Ames, Iowa, pp. 89–91.

CAST (2004) *Management of Pest Resistance: Strategies Using Crop Management Biotechnology, and Pesticides*. CAST Special Publication No. 24. Council for Agricultural Sciences and Technologies, Ames, Iowa.

Cauquil, J. (1985) La protection des cottoniers contre les ravageurs en Afrique francophone au sud du Sahara: principe et évolution des techniques. *Coton et Fibres Tropical* 4, 187–191.

Denholm, I. and Jespersen, J.B. (1997) Insecticide resistance management in Europe: recent developments and prospects. *Pesticide Science* 52, 193–195.

Denholm, I. and Rowland, M.W. (1992) Tactics for managing pesticide resistance in arthropods: theory and practice. *Annual Reviews of Entomology* 37, 91–112.

Dennehy, T.J. and Denholm, I. (1998) Goals, achievements and future challenges of the Arizona whitefly resistance management program. In: *Proceedings of the 1998 Beltwide Cotton Production Research Conference*. National Cotton Council of America, Memphis, Tennessee, pp. 68–72.

Dennehy, T.J. and Williams, L. (1997) Management of resistance in *Bemisia* in Arizona cotton. *Pesticide Science* 51, 398–406.

Dover, M.J. and Croft, B.A. (1986) Pesticide resistance and public policy. *Bioscience* 36, 78–85.

EPPO (2007a) EPPO Resistance Panel on Plant Protection Products. http://www.eppo.org/ ABOUT_EPPO/panel_composition/ppp/resistance.htm (accessed November 2007).

EPPO (2007b) *Ad hoc* EPPO Workshop on insecticide resistance of *Meligethes* spp. (pollen beetle) on oilseed rape. BBA, Berlin-Dahlem, 2007-09-03/05. Conclusions and Recommendations. http://archives.eppo.org/MEETINGS/2007_meetings/meligethes/ conclusions_meligethes.pdf (accessed December 2007).

European Commission (1993) Commission Directive (93/71/EEC) amending Council Directive 91/414/EEC concerning the placing of plant protection products on the market. *Official Journal of the European Community* L221, 31/08/1993, 27–36.

European Commission (2001) Document SANCO/2692/2001 of 25 July 2001. Working Document of the Commission Services Technical Annex to Report from the Commission to the European Parliament and the Council on the Evaluation of the Active Substances of Plant Protection Products, p. 39. http://europa.eu.int/comm/food/fs/ph_ps/pro/ppp01_ ann_en.pdf (accessed November 2007).

European Commission (2003) Regulation (EC) No 1336/2003 of 25 July amending Regulation (EC) No 2076/2002 as regards the continued use of the substances listed in Annex II (Text with EEA relevance). *Official Journal of the European Community* L187, 26/07/2003, 21–25; available at http://europa.eu.int/eur-lex/pri/en/oj/dat/2003/l_187/l_ 18720030726en00210025.pdf

European Commission (2004) 2004/129/EC: Commission Decision of 30 January concerning the non-inclusion of certain active substances in Annex I to Council Directive 91/414/EEC and the withdrawal of authorisations for plant protection products containing these substances (Text with EEA relevance) (notified under document number C(2004) 152). *Official Journal of the European Community* L037, 10/02/2004, 27-31; available at http://europa.eu.int/eur-lex/pri/en/oj/dat/2004/l_037/l_03720040210en00270031.pdf

Farrell, T (2006) Insecticide resistance management strategy (IRMS). In: *Cotton Pest Management Guide 2006–07*. State of New South Wales, NSW Agriculture, Sydney, Australia, pp. 32–50.

Fitt, G.P. (2003) Implementation and impact of transgenic *Bt* cottons in Australia. In: *Cotton Production for the New Millennium: Proceedings of the 3rd World Cotton Research Conference*. Agricultural Research Council – Institute for Industrial Crops, Pretoria, South Africa, pp. 371–381.

Forrester, N.W., Cahill, M., Bird, L.J. and Layland, J.K. (1993) Management of pyrethroid and endosulfan resistance in *Helicoverpa armigera* (Lepidoptera: Noctuidae) in Australia. *Bulletin of Entomological Research (Supplement)* 1, 1–132.

Georghiou, G.P. and Lagunes-Tejeda, A. (1991) *The Occurrence of Resistance to Pesticides in Arthropods*. Food and Agriculture Organization of the United Nations, Rome.

Glaser, J.A. and Matten, S.R. (2003) Sustainability of insect resistance management strategies for transgenic *Bt* corn. *Biotechnology Advances* 22, 45–69.

Gould, F. (1998a) Evolutionary biology and genetically engineered crops. *BioScience* 38, 26–33.

Gould, F. (1998b) Sustainability of transgenic insecticidal cultivars: integrating pest genetics and ecology. *Annual Reviews of Entomology* 43, 701–726.

Griffiths, W.T. (1984) A review of the development of cotton pest problems in the Sudan Gezira. MSc. thesis, University of London, London.

Hawkins, L.S. (1986) The role of regulatory agencies in dealing with pesticide resistance. In: *Pesticide Resistance; Strategies and Tactics for Management*. National Academy Press, Washington, DC, pp. 393–403.

Health Canada (1999) Voluntary Pesticide Resistance-Management Labeling Based on Target Site/Mode of Action, DIR99–06. http://www.hc-sc.gc.ca/pmra-arla/English/pdf/dir/ dir9906-e.pdf (accessed November 2007).

Jackson, G.J. (1986) Insecticide resistance: what is industry doing about it? In: *Proceedings of the 1986 British Crop Protection Conference – Pests and Diseases*, Vol. 3. BCPC, Alton, UK, pp. 943–949.

Johnson, E.L. (1986) Pesticide resistance management: an ex-regulator's view. In: *Pesticide Resistance; Strategies and Tactics for Management*. National Academy Press, Washington, DC, pp. 392–402.

Kranthi, K.R., Kranthi, S., Banerjee, S.K. and Mayee, C.D. (2003) Perspectives on resistance management strategies for *Bt* cotton in India. In: *Cotton Production for the New Millennium: Proceedings of the 3rd World Cotton Research Conference*. Agricultural Research Council – Institute for Industrial Crops, Pretoria, South Africa, pp. 1254–1262.

MacDonald, O.C., Meakin, I. and Richardson, D.M. (2003) The role and impact of the regulator in resistance management. In: *Proceedings of the 2003 BCPC International Congress – Crop Science and Technology*, Vol. 2. BCPC, Alton, UK, pp. 703–708.

Matten, S.R. and Beauchamp, P. (2004) The Environmental Protection Agency and the Pest Management Regulatory Agency: pest resistance management goals and challenges. In: *Management of Pest Resistance: Strategies Using Crop Management, Biotechnology, and Pesticides*. CAST Special Publication No. 24. Council for Agricultural Sciences and Technologies, Ames, Iowa, pp. 102–106.

Matten, S.R. and Reynolds, A.H. (2003) Current resistance management requirements in *Bt* cotton in the United States. *Journal of New Seed* 5, 137–178.

Matten, S.R., Hellmich, R.L. and Reynolds, A.H. (2004) Current resistance management strategies for *Bt* corn in the United States. In: Koul, O. and Dhaliwal, G.S. (eds) *Transgenic Crop Production: Concepts and Strategies*. Oxford & IBH Publishing Co. Pvt. Ltd, New Delhi, India, pp. 261–288.

National Research Council (1986) *Pesticide Resistance: Strategies and Tactics for Management*. National Academy Press, Washington, DC.

Plapp, F.W. Jr, Browning, C.R. and Sharpe, P.J.H. (1979) Analysis of rate of development of insecticide resistance based on simulation of a genetic model. *Environmental Entomology* 8, 494–500.

Roush, R.T. (1989) Designing resistance management programs: how can we choose? *Pesticide Science* 26, 423–441.

Roush, R.T. (1997a) Managing resistance to transgenic crops. In: Carozzi, N. and Koziel, M. (eds) *Advances in Insect Control: The Role of Transgenic Plants*. Taylor & Francis, London, pp. 271–294.

Roush, R.T. (1997b) *Bt*-transgenic crops: just another pretty insecticide or a chance for a new start in resistance management? *Pesticide Science* 61, 328–334.

Russell, D.A., Kranthi, K.R., Mayee, C.D., Banerjee, S.K. and Raj, S. (2003) Area-wide management of insecticide resistant pests of cotton in India. In: *Cotton Production for the New Millennium: Proceedings of the 3rd World Cotton Research Conference*. Agricultural Research Council – Institute for Industrial Crops, Pretoria, South Africa, pp. 1204–1213.

Sawicki, R.M. and Denholm, I. (1987) Management of resistance to pesticides in cotton pests. *Tropical Pest Management* 33, 262–272.

Scientific Advisory Panel (1998) Subpanel on *Bacillus thuringiensis* (*Bt*) Plant-Pesticides (February 9–10, 1998). Transmittal of the final report of the FIFRA Scientific Advisory Panel Subpanel on *Bacillus thuringiensis* (*Bt*) Plant-Pesticides and Resistance Management. Report dated April 28, 1998. Docket Number: OPPTS-00231. http://www.epa.gov/scipoly/sap/1998/february/finalfeb.pdf (accessed 1998).

Scientific Advisory Panel (2001) Subpanel on Insect Resistance Management (October 18–20, 2000). Report: Sets of scientific issues being considered by the Environmental Protection Agency regarding: *Bt* plant-pesticides risk and benefit assessments. Report dated March 12, 2001, pp. 5–33. http://www.epa.gov/scipoly/sap/2000/october/octoberfinal.pdf (accessed 2001).

Tabashnik, B.E., Carrière, Y., Dennehy, T.J., Morin, S., Sisterson, M.S., Roush, R.T., Shelton, A.M. and Zhao, J.Z. (2003) Insect resistance to transgenic *Bt* crops: lessons from the laboratory and the field. *Journal of Economic Entomology* 96, 1031–1038.

Thompson, G.D. and Head, G. (2001) Implication of regulating insect resistance management. *American Entomologist* (Spring), 6–10.

USEPA (1998) *The Environmental Protection Agency's White Paper on* Bt *Plant-Pesticide Resistance Management. US EPA, Biopesticides and Pollution Prevention Division (7511C) 14 January 1998*. EPA Publication 739-S-98-001. Environmental Protection Agency, Washington, DC.

USEPA (2001a) Biopesticides Registration Action Document: *Bacillus thuringiensis* Plant-Incorporated Protectants (10/16/01). http://www.epa.gov/pesticides/biopesticides/pips/bt_brad.htm (accessed November 2007).

USEPA (2001b) Guidance for Pesticide Registrants on Pesticide Resistance Management Labeling (PR Notice 2001-5). http://www.epa.gov/PR_Notices/pr2001-5.pdf (accessed November 2007).

Wilson, A.G.L. (1974) Resistance of *Heliothis armigera* to insecticides in the Ord Irrigation Area, North Western Australia. *Journal of Economic Entomology* 67, 256–258.

Index